CYTOGENETICS
AND
PLANT BREEDING

CYTOGENETICS AND PLANT BREEDING

Dr. Shailendra Sadashiv Choudhari

Oxford Book Company
Jaipur • New Delhi

Published By

Oxford Book Company

Regd. Off. "Bony Residency", Gate No. 2
Opp. Tilak Public School, Vishweswariya Nagar,
Gopalpura Road, Jaipur - 302018, India
Ph: 0141 - 2761280, Telefax: 2761381
email - oxfordbookjpr@gmail.com
website - oxfordbookcompany.com

Delhi Office: 102, 1st Floor, Satyam House,
4327/3, Ansari Road, Darya Ganj,
New Delhi - 110002, India
Ph: 011-45652440

Distributed by
KEYA INTERNATIONAL PRESS LLP
E-mail: keyainternationalpressllp@gmail.com

First published in India, 2024

ISBN : 9789355241146

Typeset by Shivangi Computers, Jaipur

Printed and bound in India by Replika Press Pvt. Ltd.

Contents

Dedicated to

Late Choudhari Saheb,
Seed King of Agriculture Industry in
Sambhajinagar-Maharashtra State for his
pertinent efforts in Pride of Agriculturists

Prefae

Cytogenetics is essentially a branch of genetics, but is also a part of cell biology/cytology (a subdivision of human anatomy), that is concerned with how the chromosomes relate to cell behaviour, particularly to their behaviour during mitosis and meiosis.[1] Techniques used include karyotyping, analysis of G-banded chromosomes, other cytogenetic banding techniques, as well as molecular cytogenetics such as fluorescence *in situ* hybridization (FISH) and comparative genomic hybridization (CGH).

Chromosomes were first observed in plant cells by Carl Nägeli in 1842. Their behavior in animal (salamander) cells was described by Walther Flemming, the discoverer of mitosis, in 1882. The name was coined by another German anatomist, von Waldeyer in 1888.

The next stage took place after the development of genetics in the early 20th century, when it was appreciated that the set of chromosomes (the karyotype) was the carrier of the genes. Levitsky seems to have been the first to define the karyotype as the phenotypic appearance of the somatic chromosomes, in contrast to their genic contents. Investigation into the human karyotype took many years to settle the most basic question: how many chromosomes does a normal diploid human cell contain? In 1912, Hans von Winiwarter reported 47 chromosomes in spermatogonia and 48 in oogonia, concluding an XX/XO sex determination mechanism. Painter in 1922 was not certain whether the diploid number of humans was 46 or 48, at first favoring 46. He revised his opinion later from 46 to 48, and he correctly insisted on humans having an XX/XY system of sex-determination. Considering their techniques, these results were quite remarkable. In science books, the number of human chromosomes remained at 48 for over thirty years. New techniques were needed to correct this error.

The lily is a favored organism for the cytological examination of meiosis since the chromosomes are large and each morphological stage of meiosis can be easily identified microscopically. Hotta, Chandley et al. presented the evidence for a common pattern of DNA nicking and repair synthesis in male meiotic cells of lilies and rodents during the zygotene–pachytene stages of meiosis when crossing over was presumed to occur. The presence of a common pattern between organisms as phylogenetically distant as lily and mouse led the authors to conclude that the organization for meiotic crossing-over in at least higher eukaryotes is probably universal in distribution.

Plant breeding has been very successful in developing improved varieties using conventional tools and methodologies. Nowadays, the availability of genomic tools and resources is leading to a new revolution of plant breeding, as they facilitate the study of the genotype and its relationship with the phenotype, in particular for complex traits. Next Generation Sequencing (NGS) technologies are allowing the mass sequencing of genomes and transcriptomes, which is producing a vast array of genomic information. The analysis of NGS data by means of bioinformatics developments allows discovering new genes and regulatory sequences and their positions, and makes available large collections of molecular markers. Genome-wide expression studies provide breeders with an understanding of the molecular basis of complex traits.

Cytogenetics plays an important role in understanding the chromosomal and genetic architecture of plant species.

This comprehensive book on Plant Cytogenetics presents theoretical and practical aspects of plant cytogenetics. The book presents to the reader the crucial role of cytogenetics in improving crops.

— ***Dr. Shailendra Sadashiv Choudhari***

Chapter 1

History and Scope

History of the Problem, Genetics and Epigenetics

The origins of genetics are to be found in Gregor Mendel's memoir on plant hybridization (1865). However, the word 'genetics' was only coined in 1906, to designate the new science of heredity. Founded upon the Mendelian method for analyzing the products of crosses, this science is distinguished by its explicit purpose of being a general 'science of heredity', and by the introduction of totally new biological concepts (in particular those of gene, genotype, and phenotype). In the 1910s, Mendelian genetics fused with the chromosomal theory of inheritance, giving rise to what is still called 'classical genetics'. Within this framework, the gene is simultaneously a unit of function and transmission, a unit of recombination, and of mutation. Until the early 1950s, these concepts of the gene coincided. But when DNA was found to be the material basis of inheritance, this congruence dissolved.

Historical background

Ancient theories of pangenesis and blood in heredity

Although scientific evidence for patterns of genetic inheritance did not appear until Mendel's work, history shows that humankind must have been interested in heredity long before the dawn of civilization. Curiosity must first have been based on human family resemblances, such as similarity in body structure, voice, gait, and gestures. Such notions were instrumental in the establishment of family and royal dynasties. Early nomadic tribes were interested in the qualities of the animals that they herded and domesticated and, undoubtedly, bred selectively. The first human settlements that practiced farming appear to have selected crop plants with favourable qualities. Ancient tomb paintings show racehorse breeding pedigrees containing clear depictions of the inheritance of several distinct physical traits in the horses. Despite this interest, the first recorded speculations on heredity did not exist until the time of the ancient Greeks; some aspects of their ideas are still considered relevant today.

Hippocrates (*c.* 460–*c.* 375 BCE), known as the father of medicine, believed in the inheritance of acquired characteristics, and, to account for this, he devised the hypothesis known as pangenesis. He postulated that all organs of the body of a parent gave off invisible "seeds," which were like miniaturized building components and were transmitted during sexual intercourse, reassembling themselves in the mother's womb to form a baby.

Aristotle (384–322 BCE) emphasized the importance of blood in heredity. He thought that the blood supplied generative material for building all parts of the adult body, and he reasoned that blood was the basis for passing on this generative power to the next generation. In fact, he believed that the male's semen was purified blood and that a woman's menstrual blood was her equivalent of semen. These male and female contributions united in the womb to produce a baby. The blood contained some type of hereditary essences, but he believed that the baby would develop under the influence of these essences, rather than being built from the essences themselves.

Aristotle's ideas about the role of blood in procreation were probably the origin of the still prevalent notion that somehow the blood is involved in heredity. Today people still speak of certain traits as being "in the blood" and of "blood lines" and "blood ties." The Greek model of inheritance, in which a teeming multitude of substances was invoked, differed from that of the Mendelian model. Mendel's idea was that distinct differences between individuals are determined by differences in single yet powerful hereditary factors. These single hereditary factors were identified as genes. Copies of genes are transmitted through sperm and egg and guide the development of the offspring. Genes are also responsible for reproducing the distinct features of both parents that are visible in their children.

Preformation and natural selection

In the two millennia between the lives of Aristotle and Mendel, few new ideas were recorded on the nature of heredity. In the 17th and 18th centuries the idea of preformation was introduced. Scientists using the newly developed microscopes imagined that they could see miniature replicas of human beings inside sperm heads. French biologist Jean-Baptiste Lamarck invoked the idea of "the inheritance of acquired characters," not as an explanation for heredity but as a model for evolution. He lived at a time when the fixity of species was taken for granted, yet he maintained that this fixity was only found in a constant environment. He enunciated the law of use and disuse, which states that when certain organs become specially developed as a result of some environmental need, then that state of development is hereditary and can be passed on to progeny. He believed that in this way, over many generations, giraffes could arise from deerlike animals that had to keep stretching their necks to reach high leaves on trees.

British naturalist Alfred Russel Wallace originally postulated the theory of evolution by natural selection. However, Charles Darwin's observations during his circumnavigation of the globe aboard the HMS *Beagle* (1831–36) provided evidence for natural selection and his suggestion that humans and animals shared a common ancestry. Many scientists at the time believed in a hereditary mechanism that was a version of the ancient Greek idea of pangenesis, and Darwin's ideas did not appear to fit with the theory of heredity that sprang from the experiments of Mendel.

The work of Mendel

Before Gregor Mendel, theories for a hereditary mechanism were based largely on logic and speculation, not on experimentation. In his monastery garden, Mendel carried out a large number of cross-pollination experiments between variants of the garden pea, which he obtained as pure-breeding lines. He crossed peas with yellow seeds to those with green seeds and observed that the progeny seeds (the first generation, F_1) were all yellow. When the F_1 individuals were self-pollinated or crossed among themselves, their progeny (F_2) showed a ratio of 3:1 (3/4 yellow and 1/4 green). He deduced that, since the F_2 generation contained some green individuals, the determinants of greenness must have been present in the F_1 generation, although they were not expressed because yellow is dominant over green. From the precise mathematical 3:1 ratio (of which he found several other examples), he deduced not only the existence of discrete hereditary units (genes) but also that the units were present in pairs in the pea plant and that the pairs separated during gamete formation. Hence, the two original lines of pea plants were proposed to be *YY* (yellow) and *yy* (green). The gametes from these were *Y* and *y*, thereby producing an F_1 generation of *Yy* that were yellow in colour because of the dominance of *Y*. In the F_1 generation, half the gametes were *Y* and the other half were *y*, making the F_2 generation produced from random mating 1/4 *Yy*, 1/2 *YY*, and 1/4 *yy*, thus explaining the 3:1 ratio. The forms of the pea colour genes, *Y* and *y*, are called alleles.

Mendel also analyzed pure lines that differed in pairs of characters, such as seed colour (yellow versus green) and seed shape (round versus wrinkled). The cross of yellow round seeds with green wrinkled seeds resulted in an F_1 generation that were all yellow and round, revealing the dominance of the yellow and round traits. However, the F_2 generation produced by self-pollination of F_1 plants showed a ratio of 9:3:3:1 (9/16 yellow round, 3/16 yellow wrinkled, 3/16 green round, and 1/16 green wrinkled; note that a 9:3:3:1 ratio is simply two 3:1 ratios combined). From this result and others like it, he deduced the independent assortment of separate gene pairs at gamete formation.

Mendel's success can be attributed in part to his classic experimental approach. He chose his experimental organism well and performed many controlled experiments to collect data. From his results, he developed brilliant explanatory hypotheses and went on to test these hypotheses experimentally. Mendel's methodology established a prototype for genetics that is still used today for gene discovery and understanding the genetic properties of inheritance.

Epigenetics Phenomenon

Epigenetics deals with changes in gene expression not resulting directly from mutations of DNA sequences, which lead to the formation of inherited traits both intra-generationally and inter-generationally. Let's imagine a beehive. Within a hive, all bees are genetically identical. So how do worker bees, soldier bees and queen bees come into being from the same genome? The answer lies in the diet: larvae receiving pollen develop into worker bees, and those who receive royal jelly become queens. How is it possible?

Here is the need to explain what makes up the epigenetic phenomenon. The most fascinating mechanism of epigenetics is DNA methylation, i. e. attaching a small "marker" directly to nucleotides (components – A, T, C and G) within the DNA. Histone modifications are another very important mechanism in epigenetics. But what are hystones? These are proteins that help to maintain proper spatial structure of our DNA. Adding a "marker" to the histone will locally loosen the DNA structure and thus increase the possibility of using this part of the molecule.

Can We Inherit Our Parents Experience?

In 2007 the breakthrough in the field of epigenetics surprised the world of biology. Mice with a gene called ***Agouti***, responsible for yellow colour and diabetes, causing obesity and rapid death, were fed with food as a source of methyl groups ("markers") during the experiment. To great surprise, the born offspring were brown, slim, did not suffer from diabetes and lived for a very long time. The ***Agouti*** gene still existed in their cells, but was silenced by the aforementioned methylation. What's more, the blockade appeared even in the next generation of mice. One conclusion was reached – we are not only what our parents ate, but also what our grandparents ate. Following such a trail, it turns out that apart from food, it is of great importance for the development of future generations whether we smoke, move frequently or even think positively. All these actions influence the changes in our epigenome. Today we know that DNA is not our destiny, although it plays a huge role in shaping it. Genetics loads weapons, but it is epigenetics that pulls the trigger.

Epigenetics and Diabetes

Diabetes is a chronic disease caused by hereditary and/or acquired insulin deficiency produced by the pancreas, which leads to an increase in blood glucose levels (hyperglycaemia) and thus to many complications. Chronic hyperglycaemia is associated with damage, dysfunction and failure of various organs, especially eyes, kidneys, nerves, heart and blood vessels. Diabetes research is the apple of many scientists eye around the world. Currently, more and more attention is being paid to the assessment of the influence of non-genetic factors on the development of this civilization disease. In order to demonstrate the relationship between environmental and genetic factors, studies were carried out in newborns with low birth weight. It was then hypothesized that malnutrition of the mother may be the cause of later development of diabetes in children. The results of these studies confirmed that disorders in the intrauterine environment contribute to the formation of epigenetic modifications and affect the development of the fetus. More importantly, they may lead to many diseases- including diabetes – in the future (their further development), even many years later.

Every day there are new reports on the role of epigenetic factors (especially diet and physical activity) in the increase in the occurrence of civilization diseases such as the aforementioned diabetes and cancer or very common diseases of the vascular system. In addition to the above mentioned, there is an upward trend in studies showing the impact of environmental pollution.

So far, scientists from the Foundation of Research and Science Development have managed to prove that the activity of individual genes varies depending on the method of treating diabetes mellitus. Therefore, Magdalena Gomó³ka is currently preparing a plan for the continuation of research aimed at assessing the influence of non-genetic factors and the activity of genes on the development and course of various types of diabetes.

Ideas of HEREDITY - A Brief History

Long before anyone discovered DNA, chromosomes, or genes, humans had figured out something important about the world around us. When we turned from hunting and gathering our food to growing and raising it, we realized that we could breed plants and animals to bring out desirable traits. Wild grasses became wheat under human cultivation; we domesticated and bred cows, pigs, and sheep to produce food, leather, and wool. And we also noted similarities between parents and offspring without knowing what causes these similarities. The concept of heredity preceded the science of genetics, but it is genetics that explains why all living species carry inherited traits and qualities. This timeline lists some of the key advances in thinking about heredity and genetics.

Heredity Definition

In the simplest of words, heredity refers to the passing of traits or characteristics through genes from one generation (parent) to the other generation (offspring). Heredity is very evidently seen in sexual reproduction. This is because, in this process, the variation of inherited characteristics is high.

The variation occurs due to some errors in DNA copying. Variation is important because it contributes to the evolution and forms the basis of heredity. Variation is caused due to gene mutations, the interaction of genes with the environment, and various combinations of genetic material. Remember that variation can occur through the asexual reproduction process too. But, these variations are not very noticeable.

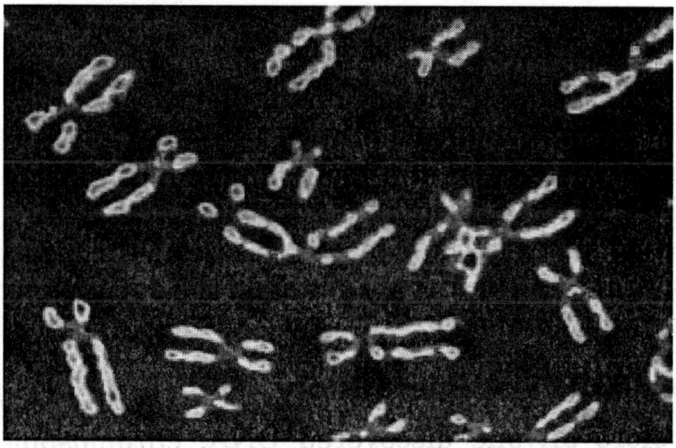

Gregor Mendel- The Father of Genetics

Acquiring characteristics or traits from one generation to the other is nothing but inheritance. Here, both parents contribute equally to the inheritance of traits. Gregor Mendel, also known as the Father of Genetics, conducted immense research and studies on this inheritance of traits.

He researched on plant breeding and hybridization and conducted his experiments on pea plants to show the inheritance of traits in living organisms.

Know Some Terms

- **Gene** – It is the basic unit of inheritance. It consists of a sequence of DNA, which is the genetic material. Genes can mutate and can take two or more alternative forms.
- **Alleles** – The alternative forms of genes. They affect the same characteristics or traits in alternate forms. They are located on the same place of the chromosome.
- **Chromosomes** – These are thread-like structures made up of nucleic acids (DNA) and proteins. They are mostly found in the nucleus of the cells. They carry the hereditary or genetic information in the form of genes.
- **Genotype** – It is the complete heritable genetic identity of an organism. It is the set of alleles that are carried by the organism. It also includes non-expressed alleles.
- ***Phenotype*** – It is the description of the actual physical characteristics of an organism or the expressed form of the genotype.

- **Dominant alleles** – When an allele affects the phenotype of an organism, then it is a dominant allele. Capital letters represent dominant alleles. For example, "T" to express tallness.
- **Recessive alleles** – An allele that affects the genotype in the absence of the dominant allele is called a recessive allele. Small letters represent recessive alleles. For example – "t" for tallness.
- **Homozygous** – Each organism has two alleles for every gene (Each chromosome has one each). In homozygous, both the alleles are same. For Example, "TT" is the homozygous expression for tallness trait.
- **Heterozygous** – If the two alleles are different from each other, then they are heterozygous in nature. For Example, "Tt" is the heterozygous expression for tallness trait.

Mendel's Experiments

Monohybrid Cross

It is the cross between two plants which have one pair of contrasting characters. For Example, a cross between a tall pea plant and a short (dwarf) plant. The following diagram explains this in detail.

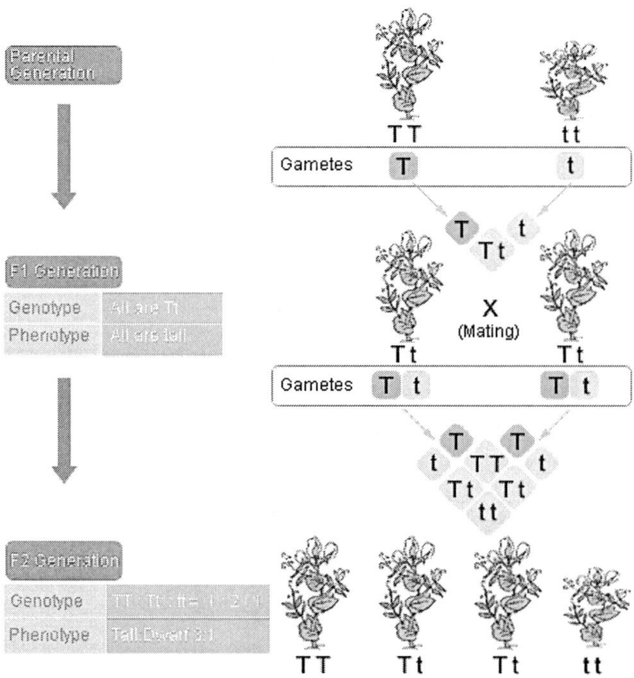

Observations & Conclusion

- In the first generation (F1), the progeny were tall. There was no medium height plant.
- In the second generation (F2), $1/4^{th}$ of the offspring were short and 3D4 were tall.
- The Phenotypic ratio in F2 – 3: 1 (3 tall: 1 short)
- The Genotypic ratio in F2 – 1: 2: 1 – (TT: Tt: tt)
- For a plant to be tall, a single copy of "T" is enough. But if a plant has to be short, both the copies should be "t"
- In Tt, 'T' is expressed and 't' is suppressed. Hence, the characters 'T' is the dominant trait and 't' is the recessive trait.

Genotype, phenotype Concept

The meaning of the terms given at the start of the introduction may at first seem clear, but conceptual questions have accompanied or been implied by the use of the terms since their introduction to English-language readers by Johannsen's (1911) "The Genotype Conception of Heredity" and up until the present. How does an individual organism's DNA influence the processes of development of its traits over its lifetime—processes that also involve other influences? How can an individual organism's traits be used to identify its DNA sequence? Why are the terms genotype and phenotype still used if they simply refer to the individual's DNA and traits? One answer to this last question is that what counts is less the meaning of the terms than what their use has come to *signify*, in particular, that certain issues have been resolved: the barrier between the organism's life course and DNA transmitted to the next generation; evolution is change in frequencies of genes or DNA sequences in populations over time; development of traits will eventually be understood in terms of a composite of the effects of DNA on the organism; what counts is what is underneath or inside the observable surface; and findings about heredity derived using one meaning of terms affirm findings derived using a different meaning. How firmly resolved are these issues? If one wants to open them up for further scrutiny, it could be helpful to question the meaning of genotype, phenotype, and the distinction between them.

Another response to why the terms continue to be used is that the predominant meanings are not the only ones. Type connotes a *class*. This sense of the terms leads to further questions: What makes organisms similar enough to be grouped in a genotype as a class? ...or in a phenotype as a class? How is the membership of organisms in a genotype manifest in their membership in a phenotype—or phenotypes? How can the membership of organisms in a phenotype be used to identify their membership in a genotype? Type also connotes an *abstraction* away from the full set of observed characteristics. What then is to be accentuated and what de-emphasized about a genotype and phenotype, as DNA and traits or as classes—especially when asking any of the previous questions? And how—by what concepts, methods, and models—is what has been de-emphasized to be brought back and *re-integrated* into the scientific account?

To make sense of the original meanings of the terms and the distinction between them as well as the coexistence of different kinds of meanings since then, a specific kind of abstraction is pertinent, namely, the *material* practices of *control* over biological materials and conditions advanced in modern experimental biology and agricultural breeding or entailed in the allied use of *models in analysis of data*. Reliable methods more than endorsable theories are the touchstone of this kind of abstraction. It is within a framing of control—and the consequent challenge of reintegration—that this entry discusses the other senses already mentioned of the terms genotype and phenotype: in the context of the relationship between an individual's DNA and traits; what the use of the terms signifies; and grouping into classes. (Given the continuing coexistence of the different kinds of meaning of the terms, qualifiers are used when there might be ambiguity.)

Although there may be grounds to revise various positions and inquiries about the genotype-phenotype *relationship* in light of assumptions made about control and eventual reintegration, such discussion lies beyond the scope of this entry. The entry also cannot do justice to the rich social and economic history of heredity, where concrete cases abound concerning the control of biological materials for production and breeding—dating from well before the genotype-phenotype distinction through to present-day biotechnology.

The norm of reaction Phenocopies

The **Norm of Reaction** is a curve that relates, for a given **genotype**, the contribution of **environmental variation** to observed **phenotypic variation**. The shape of the curve may be essentially flat across environments [***Left***], such that the expected phenotype of any given genotype is highly predictable, independent of the environment. In this model, the more narrow the distribution, the less variable the trait. This is what is typically meant when a trait is said to be "*genetic*." In the first example, Genotype **A** always has a higher trait value than Genotype **B**, and this trait value is quite constant across environments.

Alternatively, the Norms of Reaction for two genotypes may have parallel slopes [***Middle***], such that *relative* rankings of trait values of Genotypes **A** & **B** are always the same (**A** is superior to **B**), but the *absolute* trait value is dependent on environment. In the second example, **B** has the higher trait value in either environment, but the trait value of **B** in the "**X**" environment exceeds that of **A** in the "**Y**" environment. The **heritability** of the trait would be close to **1.0** in either environment, but this does not predict the behavior in the other environment.

Finally, the shape of the Norm of Reaction for two different genotypes may be quite different, such that it is not possible to predict from analysis of one genotype the response of the another. Consider two breeds of cattle (**genotypes A** & **B**). Among cattle raised in a free-range, nutrient-poor environment (environmental "**Y**"), those with genotype **A** always produce more milk fat (narrow red curve) than do those with genotype **B** (broader blue curve). The range of phenotypes within each breed is relatively constant over a wide range of free-range environments. A conventional study of the heritability of milkfat production conducted in this environment will conclude that genes have a strong influence on milkfat production (**H** ~ 1.0), that is, the difference between breeds is "***mostly genetic***," as in the middle example.

On the other hand, when the same two breeds are moved to a feed-lot environment with abundant nutrients (environment "**X**"), both breeds show a marked improvement in average milkfat production. Further, cattle with genotype **B** now typically produce richer milk than those with genotype **A** (mean of blue curve slightly higher), *the reverse of the previous situation.* Both breeds also show a wide range of milk fat production (broad blue and red curves), depending on the exact environmental conditions [e.g., feed types]. A study of heritability in this environment will conclude that genes have relatively low influence on milk fat production (low heritability), which is mostly a consequence of **environmental variation**.

Thus the relative importance of "**genes**" and "**environment**" is not a unitary value, and may vary greatly depending on exactly which environments the genes are expressed in. Studies of heritability carried out in a single environment cannot accurately estimate the Norm of Reaction, and often may not predict phenotypic response in a different environment.

SCOPE AND SIGNIFICANCE OF GENETICS

In the last few decades, the science of genetics has pervaded all aspects of biology so that it has assumed a central position of great significance in biology as a whole. While on the one hand, genetics is used for a study of the mechanism of heredity and variation, on the other hand it has provided tools for the study of the fundamental biological processes examined and taught in areas, like plant physiology, biochemistry, biosysterjiatics, ecology, plant pathology, microbiology, etc. Consequently today every biologist should be bit of a geneticist. Genetics, in fact provided the modern paradigm (a prototype) for whole of biology. The science of genetics also had a tremendous impact in applied areas including medicine, agriculture, forestry, fisheries, law and religion.

In view of this, all newspapers often address questions dealing with different aspect of genetics that may be of significance to common man, who is not a genetici' or a biologist. The recent upsurge of biotechnology has added further to the significance of the science of genetics, so that the products of genetics have also become a subject of discussion for Trade Related Aspects of Intellectual Properties (TRIPs) under the aegis of **General Agreement on Tariffs and Trade (GATT). Patenting** of life forms which may or may not be the product of genetic manipulation is one such topic, which is receiving considerable attention of both developed and developing countries.

Genetics can be broadly classified in the following three areas for the convenience of a discussion on its scope and significance : (i) **transmission** genetics involving study of transmission of genetic material from one generation to the other; (ii) **molecular and biochemical genetics,** involving study of the structure and function of genes and (iii) **population and biometrical genetics,** involving study of the behavior and effects of genes in population, often using mathematical models. The above classification is arbitrary, and the three areas are inter-related and even enter other areas of biology to answer some difficult questions. Significance of genetics also stems from the fact that the genetic material containing information for hereditary traits consists of nucleic acids only, across the entire spectrum of life on the earth.

More important of tlie two types of nucleic acids, **deoxyribose nucleic acid (DNA)** and **ribose nucleic acid (RNA),** is the former i.e. DNA, which has two unique properties : (i) it can replicate and produce its exact copies and (ii) it carries the genetic information, necessary to give form to an organism; this information is written into the sequence of four monomers called nucleotides, which make the polymer molecule, the DNA.

Genetics

The science of heredity or genetics is the study of two inconsistent elements of nature: heredity and variation. The process of transmission of characters from one generation to the next, either by gametes– sperms and ova– in sexual reproduction or by the asexual reproductive bodies in asexual recreation, is called inheritance or heredity.

Heredity is the reason for resemblances between individuals. This is the factor that siblings with the very same parents resemble each other and with their moms and dads. Variation is the reason for distinctions between individuals. This is the reason that brothers and sisters who do resemble each other are still unique people.

Hence, we have no trouble in acknowledging the distinctions between sisters, for example, and even identical twins are recognized as distinct individuals by their parents and friends. The science of genes tries to describe the system and the basis for both resemblances and differences between associated people. It also tries to discuss the phenomenon of evolution and cytodifferentiation.

What is Genetics?

The life science which handles the mechanism of heredity and reasons for variations in living beings (viruses, bacteria, plants, and animals) is referred to as genetics. The word **genetic** was derived from the Greek root **gen** which means to become or to grow into and it was coined by Bateson in 1906 for the research study of the physiology of heredity and variations.

Scope of Genetics

Geneticists study all aspects of genes. The study of the mode of gene transmission from generation to generation is broadly called transmission genetics; the study of structure and function of the gene forms molecular biology, and the study of behavior of genes in populations is called population genetics.

These 3 significant subdivisions of genetics are arbitrary and there is considerable overlapping. It is the knowledge of how genes act and how they are sent down through the generations that have actually merged biology; formerly, the particular set of biological phenomena had actually each been designated to separate disciplines.

An understanding of how genes act is now an essential requirement for such biological fields of study as development, cytology, physiology, and morphology. An understanding of gene transmission is a basic aspect of areas such as ecology, evolution, and taxonomy.

Further unification has aroused from the discovery that the fundamental chemistry of gene structure and function is really comparable across the entire spectrum of life on the earth. Hence, not so long earlier, biology was divided into numerous camps that seldom communicated with each other; today, nevertheless, every biologist should be a little bit of a geneticist, since the findings and strategies of genes are being applied and utilized in all fields. In fact, genetics contributed to the modern-day model for all of biology. It offers a unifying thread for the formerly diverse fields of biology.

Importance of Genetics

Like other disciplines of science, the genetical insight has produced brand-new obstacles as well as solutions to some human problems.

1. For instance, early in this century, a new wheat strain called Marquis was developed in Canada. This high quality strain is resistant to diseases; furthermore, it grows two weeks earlier than other commercially used strain– a really crucial aspect where the growing season is short. The introduction of the Marquis strain of wheat had opened countless square kilometers of fertile soil to grow in such northern countries as Canada, Sweden, and the USSR. Also, the IR26 strain of rice was developed by geneticists in 1973; it has a wide variety of desirable qualities such as resistance to a number of viral and fungal diseases and defense from insects such as green leafhopper, brown hopper, and stem borer. In addition to enhancing crop varieties, geneticists have discovered to change the genetic systems of insects to decrease their fertility. This strategy has actually supplied an essential new tool in the olden battle to keep bugs away from human crops and habitations.

2. Over the last few years, advances in biotechnology have actually resulted in the creation of special genetically engineered strains of bacteria and fungi that carry particular genes from unrelated organisms such as humans. These microbes are used to produce such useful compounds as insulin, human growth hormone, and the antiviral (or anti-cancer) agent– the interferon.

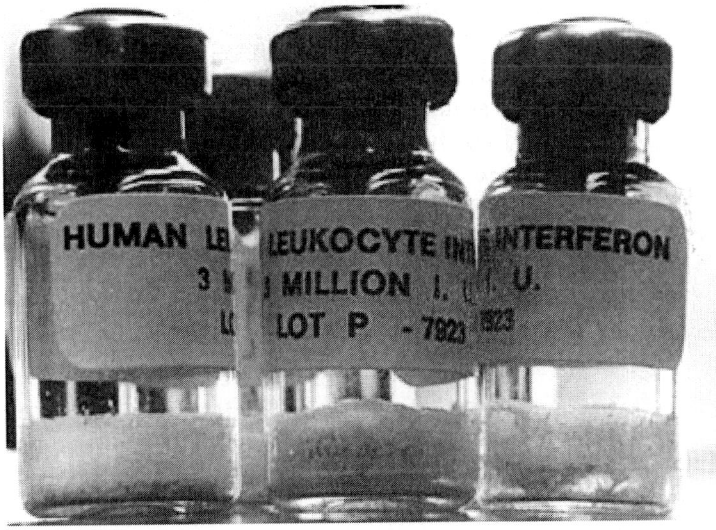

3. Our new ability to diagnose genetic disease poses a crucial ethical dilemma. An approximated 5 percent of our population survives with serious physical or mental genetic defects. This portion most likely will increase with prolonged direct exposure to numerous environmental elements and paradoxically, with better medical innovation. For example, of those patients admitted

to pediatric hospitals in North America, 30 percent approximated to have genetic diseases. This will certainly increase terrific the financial problem of human society. Thus, knowledge of medicinal and clinical genetics is very important.

4. In the modern period the genetics has changed agriculture, cultivation, horticulture, animal husbandry, and numerous other branches of science. The science of genetics has actually shown worth in removing numerous faulty concepts of the man about inheritance.

Classical Genetics

Classical genetics is the branch of genetics based solely on visible results of reproductive acts. It is the oldest discipline in the field of genetics, going back to the experiments on Mendelian inheritance by Gregor Mendel who made it possible to identify the basic mechanisms of heredity. Subsequently, these mechanisms have been studied and explained at the molecular level.

Classical genetics consists of the techniques and methodologies of genetics that were in use before the advent of molecular biology. A key discovery of classical genetics in eukaryotes was genetic linkage. The observation that some genes do not segregate independently at meiosis broke the laws of Mendelian inheritance and provided science with a way to map characteristics to a location on the chromosomes. Linkage maps are still used today, especially in breeding for plant improvement.

After the discovery of the genetic code and such tools of cloning as restriction enzymes, the avenues of investigation open to geneticists were greatly broadened. Some classical genetic ideas have been supplanted with the mechanistic understanding brought by molecular discoveries, but many remain intact and in use. Classical genetics is often contrasted with reverse genetics, and aspects of molecular biology are sometimes referred to as molecular genetics.

Basic definitions

At the base of classical genetics is the concept of a gene, the hereditary factor tied to a particular simple feature (or character).

The set of genes for one or more characters possessed by an individual is the genotype. A diploid individual often has two alleles for the determination of a character.

Overview

Classical genetics is the aspect of genetics concerned solely with the transmission of genetic traits via reproductive acts. Genetics is, generally, the study of genes, genetic variation, and heredity. The process by which characteristics are passed down from parents to their offspring is called heredity. In the sense of classical genetics, variation is known as the lack of resemblance in related individuals and can be categorized as discontinuous or continuous. Genes are a fundamental part of DNA that is aligned linearly on a eukaryotic chromosome. Chemical information that is transported and encoded by each gene is referred to as a trait. Many organisms possess two genes for each individual trait that is present within that particular individual. These paired genes that control the same trait is classified as an allele. In an individual, the allelic genes that are expressed can be either homozygous, meaning the same, or heterozygous, meaning different. Many pairs of alleles have differing effects that are portrayed in an offspring's phenotype and genotype. The phenotype is a general term that defines an individual's visible, physical traits. The genotype of an offspring is known as its genetic makeup. The alleles of genes can either be dominant or recessive. A dominant allele needs only one copy to be expressed while a recessive allele needs two copies (homozygous) in a diploid organism to be expressed. Dominant and recessive alleles help to determine the offspring's genotypes, and therefore phenotypes.

History

Classical genetics is often referred to as the oldest form of genetics, and began with Gregor Mendel's experiments that formulated and defined a fundamental biological concept known as Mendelian inheritance. Mendelian inheritance is the process in which genes and traits are passed from a set of parents to their offspring. These inherited traits are passed down mechanistically with one gene from one parent and the second gene from another parent in sexually reproducing organisms. This creates the pair of genes in diploid organisms. Gregor Mendel started his experimentation and study of inheritance with phenotypes of garden peas and continued the experiments with plants. He focused on the patterns of the traits that were being passed down from one generation to the next generation. This was assessed by test-crossing two peas of different colors and observing the resulting phenotypes. After determining how the traits were likely inherited, he began to expand the amount of traits observed and tested and eventually expanded his experimentation by increasing the number of different organisms he tested.

About 150 years ago, Gregor Mendel published his first experiments with the test crossing of *Pisum* peas. Seven different phenotypic characteristics were studied and tested in the peas, including seed color, flower color and seed shape. The seven different characteristics which Mendel selected / checked for the experiment were as follows:

- He checked the different shape of the ripen seeds
- The color of the seed's albumen was checked
- He then selected seed coat color
- Shape of the ripen pods was seen
- Color of the unripened pods was checked
- Flower position on the axial was checked
- Height of the plant was checked, as if it is tall or dwarf.

Mendel took peas that had differing phenotypic characteristics and test-crossed them to assess how the parental plants passed the traits down to their offspring. He started by crossing a round, yellow and round, green pea and observed the resulting phenotypes. The results of this experiment allowed him to see which of these two traits was dominant and which was recessive based upon the number of offspring with each phenotype. Mendel then chose to further his experiments by crossing a pea plant homozygous dominant for round and yellow phenotypes with a pea plant that was homozygous recessive for wrinkled and green. The plants that were originally crossed are known as the parental generation, or P generation, and the offspring resulting from the parental cross is known as the first filial, or F1, generation. The plants of the F1 generation resulting from this hybrid cross were all heterozygous round and yellow seeds.

Classical genetics is a hallmark of the start of great discovery in biology, and has led to increased understanding of multiple important components of molecular genetics, human genetics, medical genetics, and much more. Thus, reinforcing Mendel's nickname as the father of modern genetics.

In other words, we can say that classical genetics is basis of the modern genetics. Classical genetics is the Mendelian genetics or the older concepts of the genetics, which solely expressed based on the phenotypes resulted from breeding experiments while the modern genetics is the new concept of genetics, which allows the direct investigation of genotypes together with phenotypes.

Monohybrid Cross (3:1)

GAMETES	R	r
Y	YR	Yr
Y	yR	yr

Dihybrid Cross (9:3:3:1)

GAMETES	YR	yR	Yr	yr
YR	YYRR	YyRR	YYRr	YyRr
yR	YyRR	yyRR	YyRr	yyRr
Yr	YYRr	YyRr	YYrr	Yyrr
Yr	YyRr	yyRr	Yyrr	Yyrr

Molecular Genetics

Molecular genetics is a sub-field of biology that addresses how differences in the structures or expression of DNA molecules manifests as variation among organisms. Molecular genetics often applies an "investigative approach" to determine the structure and/or function of genes in an organism's genome using genetic screens. The field of study is based on the merging of several sub-fields in biology: classical Mendelian inheritance, cellular biology, molecular biology, biochemistry, and biotechnology. Researchers search for mutations in a gene or induce mutations in a gene to link a gene sequence to a specific phenotype. Molecular genetics is a powerful methodology for linking mutations to genetic conditions that may aid the search for treatments/ cures for various genetics diseases.

History

For molecular genetics to develop as a discipline, several scientific discoveries were necessary. The discovery of DNA as a means to transfer the genetic code of life from one cell to another and between generations was essential for identifying the molecule responsible for heredity. Molecular genetics arose initially from studies involving genetic transformation in bacteria. In 1944 Avery, McLeod and McCarthy isolated DNA from a virulent strain of *S. pneumoniae*, and using just this DNA were able to convert a harmless strain to virulence. They called the uptake, incorporation and expression of DNA by bacteria "transformation". This finding suggested that DNA is the genetic material of bacteria. Since its discovery in 1944 genetic transformation has been found to occur in numerous bacterial species including many species that are pathogenic to humans. Bacterial transformation is often induced by conditions of stress, and the function of transformation appears to be repair of genomic damage.

The phage group was an informal network of biologists centered on Max Delbrück that contributed substantially to molecular genetics and the origins of molecular biology during the period from about 1945 to 1970. The phage group took its name from bacteriophages, the bacteria-infecting viruses that the group used as experimental model organisms. Studies by molecular geneticists affiliated with this group contributed to current understanding of how gene-encoded proteins function in DNA replication, DNA repair and DNA recombination, and on how viruses are assembled from protein and nucleic acid components (molecular morphogenesis). Furthermore, the role of chain terminating codons was elucidated. One noteworthy study was performed by Sydney Brenner and collaborators using amber mutants defective in the gene encoding the major head protein of bacteriophage T4. This study demonstrated the co-linearity of the gene with its encoded polypeptide, thus providing strong evidence for the "sequence hypothesis" that the amino acid sequence of a protein is specified by the nucleotide sequence of the gene determining the protein.

Watson and Crick (in conjunction with Franklin and Wilkins) figured out the structure of DNA, a cornerstone for molecular genetics. The isolation of a restriction endonuclease in *E. coli* by Arber and Linn in 1969 opened the field of genetic engineering. Restriction enzymes were used to linearize DNA for separation by electrophoresis and Southern blotting allowed for the identification of specific DNA segments via hybridization probes. In 1971, Berg utilized restriction enzymes to create the first recombinant DNA molecule and first recombinant DNA plasmid. In 1972, Cohen and Boyer created the first recombinant DNA organism by inserting recombinant DNA plasmids into *E. coli*, now known as bacterial transformation, and paved the way for molecular cloning. The development of DNA sequencing techniques in the late

1970s, first by Maxam and Gilbert, and then by Frederick Sanger, was pivotal to molecular genetic research and enabled scientists to begin conducting genetic screens to relate genotypic sequences to phenotypes. Polymerase chain reaction (PCR) using Taq polymerase, invented by Mullis in 1985, enabled scientists to create millions of copies of a specific DNA sequence that could be used for transformation or manipulated using agarose gel separation. A decade later, the first whole genome was sequenced (*Haemophilus influenzae*), followed by the eventual sequencing of the human genome via the Human Genome Project in 2001. The culmination of all of those discoveries was a new field called genomics that links the molecular structure of a gene to the protein or RNA encoded by that segment of DNA and the functional expression of that protein within an organism. Today, through the application of molecular genetic techniques, genomics is being studied in many model organisms and data is being collected in computer databases like NCBI and Ensembl. The computer analysis and comparison of genes within and between different species is called bioinformatics, and links genetic mutations on an evolutionary scale.

The Central Dogma

This image shows an example of the central dogma using a DNA strand being transcribed then translated and showing important enzymes used in the processes

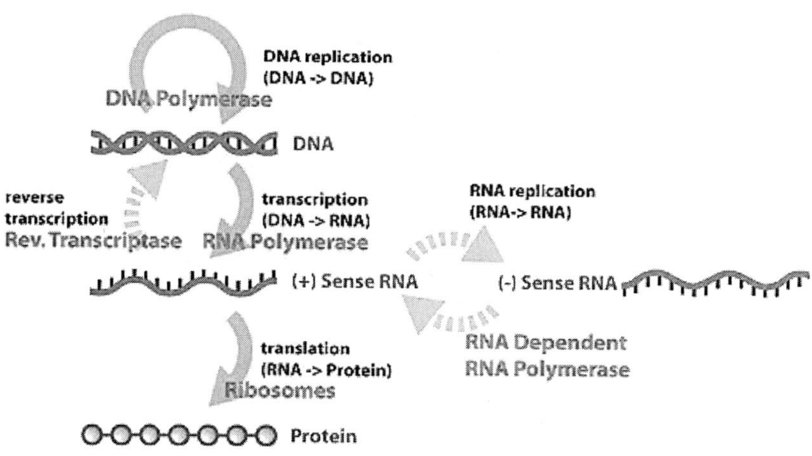

Figure: This image shows an example of the central dogma using a DNA strand being transcribed then translated and showing important enzymes used in the processes

The Central Dogma is the basis of all genetics and plays a key role in the study of molecular genetics. The Central Dogma states that DNA replicates itself, DNA is transcribed into RNA, and RNA is translated into proteins. Along with the Central Dogma, the genetic code is used in understanding how RNA is translated into proteins. Replication of DNA and transcription from DNA to mRNA occurs in the nucleus while translation from RNA to proteins occurs in the ribosome. The genetic code is made of four base pairs: adenine, cytosine, uracil, and guanine and is redundant meaning multiple combinations of these base pairs (which are read in triplicate) produce the same amino acid. Proteomics and genomics are fields in biology that come out of the study of molecular genetics and the Central Dogma.

Techniques

Forward genetics

Forward genetics is a molecular genetics technique used to identify genes or genetic mutations that produce a certain phenotype. In a genetic screen, random mutations are generated with mutagens (chemicals or radiation) or transposons and individuals are screened for the specific phenotype. Often, a secondary assay in the form of a selection may follow mutagenesis where the desired phenotype is difficult to observe, for example in bacteria or cell cultures. The cells may be transformed using a gene for antibiotic resistance or a fluorescent reporter so that the mutants with the desired phenotype are selected from the non-mutants.

Mutants exhibiting the phenotype of interest are isolated and a complementation test may be performed to determine if the phenotype results from more than one gene. The mutant genes are then characterized as dominant (resulting in a gain of function), recessive (showing a loss of function), or epistatic (the mutant gene masks the phenotype of another gene). Finally, the location and specific nature of the mutation is mapped via sequencing. Forward genetics is an unbiased approach and often leads to many unanticipated discoveries, but may be costly and time consuming. Model organisms like the nematode worm *Caenorhabditis elegans*, the fruit fly *Drosophila melanogaster*, and the zebrafish *Danio rerio* have been used successfully to study phenotypes resulting from gene mutations.

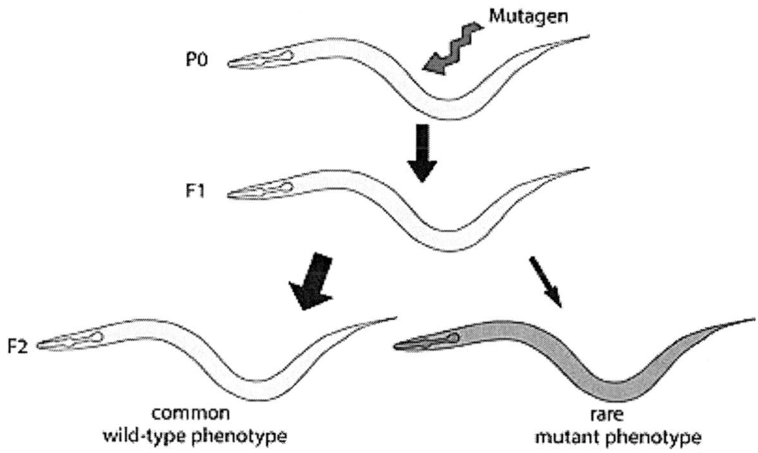

An example of forward genetics in *C. elegans* (a nematode) using mutagenesis.

Reverse genetics

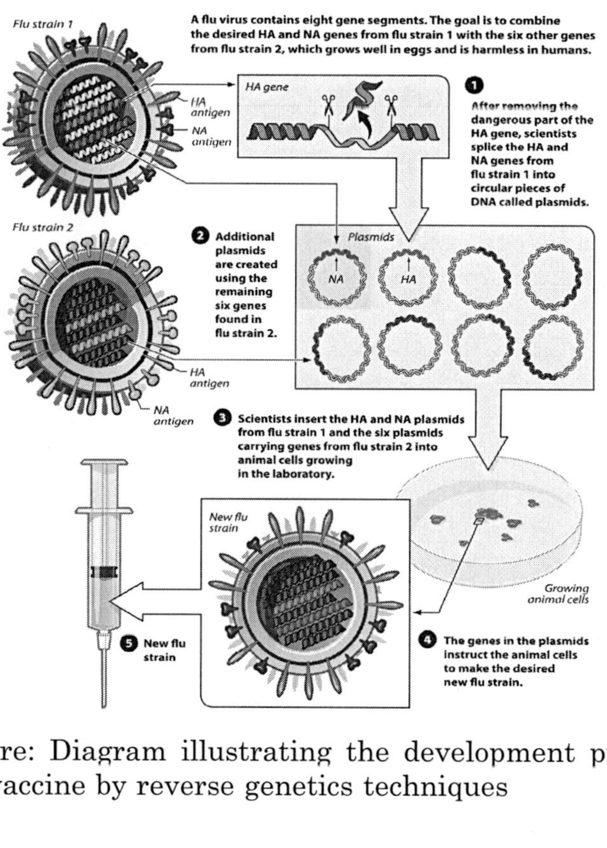

Figure: Diagram illustrating the development process of avian flu vaccine by reverse genetics techniques

Reverse genetics is the term for molecular genetics techniques used to determine the phenotype resulting from an intentional mutation in a gene of interest. The phenotype is used to deduce the function of the un-mutated version of the gene. Mutations may be random or intentional changes to the gene of interest. Mutations may be a mis-sense mutation caused by nucleotide substitution, a nucleotide addition or deletion to induce a frameshift mutation, or a complete addition/deletion of a gene or gene segment. The deletion of a particular gene creates a gene knockout where the gene is not expressed and a loss of function results (e.g. knockout mice). Mis-sense mutations may cause total loss of function or result in partial loss of function, known as a knockdown. Knockdown may also be achieved by RNA interference (RNAi). Alternatively, genes may be substituted into an organism's genome (also known as a transgene) to create a gene knock-in and result in a gain of function by the host. Although these techniques have some inherent bias regarding the decision to link a phenotype to a particular function, it is much faster in terms of production than forward genetics because the gene of interest is already known.

Recombinant DNA technology

The technology used for producing artificial DNA through the combination of different genetic materials (DNA) from different sources is referred to as Recombinant DNA Technology. Recombinant DNA technology is popularly known as genetic engineering.

The recombinant DNA technology emerged with the discovery of **restriction enzymes** in the year 1968 by Swiss microbiologist Werner Arber,

Inserting the desired gene into the genome of the host is not as easy as it sounds. It involves the selection of the desired gene for administration into the host followed by a selection of the perfect vector with which the gene has to be integrated and recombinant DNA formed.

Thus the recombinant DNA has to be introduced into the host. And at last, it has to be maintained in the host and carried forward to the offspring.

Tools Of Recombinant DNA Technology

The enzymes which include the restriction enzymes help to cut, the polymerases- help to synthesize and the ligases- help to bind. The restriction enzymes used in recombinant DNA technology play a major role in determining the location at which the desired gene is inserted into the vector genome. They are two types, namely Endonucleases and Exonucleases.

The Endonucleases cut within the DNA strand whereas the Exonucleases remove the nucleotides from the ends of the strands. The restriction endonucleases are sequence-specific which are usually palindrome sequences and cut the DNA at specific points. They scrutinize the length of DNA and make the cut at the specific site called the restriction site. This gives rise to sticky ends in the sequence. The desired genes and the vectors are cut by the same restriction enzymes to obtain the complementary sticky notes, thus making the work of the ligases easy to bind the desired gene to the vector.

The vectors – help in carrying and integrating the desired gene. These form a very important part of the tools of recombinant DNA technology as they are the ultimate vehicles that carry forward the desired gene into the host organism. Plasmids and bacteriophages are the most common vectors in recombinant DNA technology that are used as they have a very high copy number. The vectors are made up of an origin of replication- This is a sequence of nucleotides from where the replication starts, a selectable marker – constitute genes which show resistance to certain antibiotics like ampicillin; and cloning sites – the sites recognized by the restriction enzymes where desired DNAs are inserted.

Host organism – into which the recombinant DNA is introduced. The host is the ultimate tool of recombinant DNA technology which takes in the vector engineered with the desired DNA with the help of the enzymes.

There are a number of ways in which these recombinant DNAs are inserted into the host, namely – microinjection, biolistics or gene gun, alternate cooling and heating, use of calcium ions, etc.

Process of Recombinant DNA Technology

The complete process of recombinant DNA technology includes multiple steps, maintained in a specific sequence to generate the desired product.

Step-1. Isolation of Genetic Material.

The first and the initial step in Recombinant DNA technology is to isolate the desired DNA in its pure form i.e. free from other macromolecules.

Step-2. Cutting the gene at the recognition sites.

The restriction enzymes play a major role in determining the location at which the desired gene is inserted into the vector genome. These reactions are called 'restriction enzyme digestions'.

Step-3. Amplifying the gene copies through Polymerase chain reaction (PCR).

It is a process to amplify a single copy of DNA into thousands to millions of copies once the proper gene of interest has been cut using restriction enzymes.

Step-4. Ligation of DNA Molecules.

In this step of Ligation, the joining of the two pieces – a cut fragment of DNA and the vector together with the help of the enzyme DNA ligase.

Step-5. Insertion of Recombinant DNA Into Host.

In this step, the recombinant DNA is introduced into a recipient host cell. This process is termed as Transformation. Once the recombinant DNA is inserted into the host cell, it gets multiplied and is expressed in the form of the manufactured protein under optimal conditions.

As mentioned in Tools of recombinant DNA technology, there are various ways in which this can be achieved. The effectively transformed cells/organisms carry forward the recombinant gene to the offspring.

Application of Recombinant DNA Technology

- DNA technology is also used to detect the presence of HIV in a person.
- Gene Therapy – It is used as an attempt to correct the gene defects which give rise to heredity diseases.
- Clinical diagnosis – ELISA is an example where the application of recombinant
- Recombinant DNA technology is widely used in Agriculture to produce genetically-modified organisms such as Flavr Savr tomatoes, golden rice rich in proteins, and Bt-cotton to protect the plant against ball worms and a lot more.
- In the field of medicines, Recombinant DNA technology is used for the production of Insulin.

DNA Cloning

A clone is a cluster of individual entities or cells that are descended from one progenitor. Clones are genetically identical as the cell simply replicates producing identical daughter cells every time. Scientists are able to generate multiple copies of a single fragment of DNA, a gene which can be used to create identical copies constituting a DNA clone. **DNA cloning** takes place through the insertion of DNA fragments into a tiny DNA molecule. This molecule is made to replicate within a living cell, for instance, a bacterium. The tiny replicating molecule is known as the carrier of the DNA vector.

Yeast cells, viruses, and Plasmids are the most commonly used vectors. Plasmids are circular DNA molecules that are introduced from bacteria. They are not part of the main cellular genome. It carries genes, which provide the host cell with beneficial properties such as mating ability, and drug resistance. They can be conveniently manipulated as they are small enough and they are capable of carrying extra DNA which is weaved into them.

Applications Of Gene Cloning

Listed below are the applications of gene cloning:

- Gene Cloning plays an important role in the medicinal field. It is used in the production of hormones, vitamins and antibiotics.
- Gene cloning finds its applications in the agricultural field. Nitrogen fixation is carried out by cyanobacteria wherein desired genes can be used to enhance the productivity of crops and improvement of health. This practice reduces the use of fertilizers hence chemical-free produce is generated
- It can be applied to the science of identifying and detecting a clone containing a particular gene which can be manipulated by growing in a controlled environment
- It is used in gene therapy where a faulty gene is replaced by the insertion of a healthy gene. Medical ailments such as leukaemia and sickle cell anaemia can be treated with this principle.

DNA

Recombinant DNA (rDNA) molecules are DNA molecules formed by laboratory methods of genetic recombination (such as molecular cloning) that bring together genetic material from multiple sources, creating sequences that would not otherwise be found in the genome.

Recombinant DNA is the general name for a piece of DNA that has been created by combining two or more fragments from different sources. Recombinant DNA is possible because DNA molecules from all organisms share the same chemical structure, differing only in the nucleotide sequence. Recombinant DNA molecules are sometimes called **chimeric DNA** because they can be made of material from two different species like the mythical chimera. rDNA technology uses palindromic sequences and leads to the production of sticky and blunt ends.

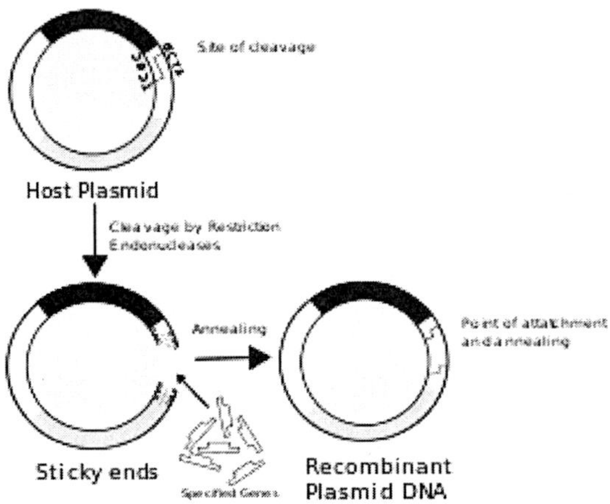

Figure: Construction of recombinant DNA, in which a foreign DNA fragment is inserted into a plasmid vector. In this example, the gene indicated by the white color is inactivated upon insertion of the foreign DNA fragment.

The DNA sequences used in the construction of recombinant DNA molecules can originate from any species. For example, plant DNA can be joined to bacterial DNA, or human DNA can be joined with fungal DNA. In addition, DNA sequences that do not occur anywhere in nature can be created by the chemical synthesis of DNA and incorporated into recombinant DNA molecules. Using recombinant DNA technology and synthetic DNA, any DNA sequence can be created and introduced into living organisms.

Proteins that can result from the expression of recombinant DNA within living cells are termed *recombinant proteins*. When recombinant DNA encoding a protein is introduced into a host organism, the recombinant protein is not necessarily produced. Expression of foreign proteins requires the use of specialized expression vectors and often necessitates significant restructuring by foreign coding sequences.

Recombinant DNA differs from genetic recombination in that the former results from artificial methods while the latter is a normal biological process that results in the remixing of existing DNA sequences in essentially all organisms.

DNA creation

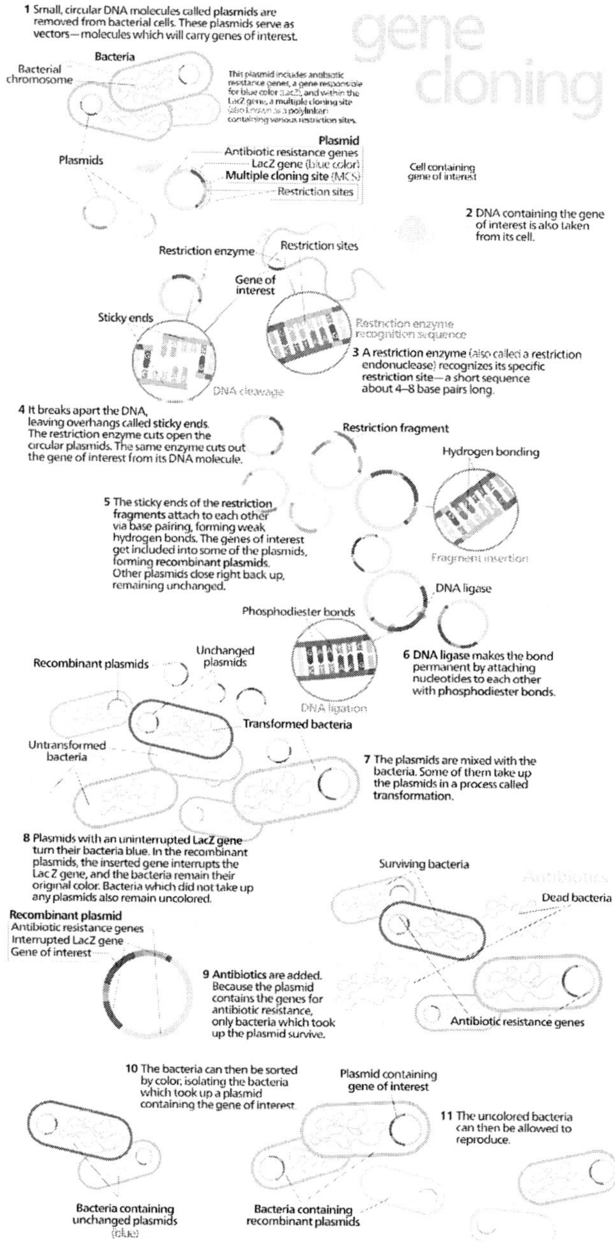

Molecular cloning is the laboratory process used to create recombinant DNA. It is one of two most widely used methods, along with polymerase chain reaction (PCR), used to direct the replication of any specific DNA sequence chosen by the experimentalist. There are two fundamental differences between the methods. One is that molecular cloning involves replication of the DNA within a living cell, while PCR replicates DNA in the test tube, free of living cells. The other difference is that cloning involves cutting and pasting DNA sequences, while PCR amplifies by copying an existing sequence.

Formation of recombinant DNA requires a cloning vector, a DNA molecule that replicates within a living cell. Vectors are generally derived from plasmids or viruses, and represent relatively small segments of DNA that contain necessary genetic signals for replication, as well as additional elements for convenience in inserting foreign DNA, identifying cells that contain recombinant DNA, and, where appropriate, expressing the foreign DNA. The choice of vector for molecular cloning depends on the choice of host organism, the size of the DNA to be cloned, and whether and how the foreign DNA is to be expressed. The DNA segments can be combined by using a variety of methods, such as restriction enzyme/ligase cloning or Gibson assembly.

In standard cloning protocols, the cloning of any DNA fragment essentially involves seven steps: (1) Choice of host organism and cloning vector, (2) Preparation of vector DNA, (3) Preparation of DNA to be cloned, (4) Creation of recombinant DNA, (5) Introduction of recombinant DNA into the host organism, (6) Selection of organisms containing recombinant DNA, and (7) Screening for clones with desired DNA inserts and biological properties.

DNA expression

Following transplantation into the host organism, the foreign DNA contained within the recombinant DNA construct may or may not be expressed. That is, the DNA may simply be replicated without expression, or it may be transcribed and translated and a recombinant protein is produced. Generally speaking, expression of a foreign gene requires restructuring the gene to include sequences that are required for producing an mRNA molecule that can be used by the host's translational apparatus (e.g. promoter, translational initiation signal, and transcriptional terminator). Specific changes to the host organism may be made to improve expression of the ectopic gene. In addition, changes may be needed to the coding sequences as well, to optimize translation, make the protein soluble, direct the recombinant protein to the proper cellular or extracellular location, and stabilize the protein from degradation.

Properties of organisms containing recombinant DNA

In most cases, organisms containing recombinant DNA have apparently normal phenotypes. That is, their appearance, behavior and metabolism are usually unchanged, and the only way to demonstrate the presence of recombinant sequences is to examine the DNA itself, typically using a polymerase chain reaction (PCR) test. Significant exceptions exist, and are discussed below.

If the rDNA sequences encode a gene that is expressed, then the presence of RNA and/or protein products of the recombinant gene can be detected, typically using RT-PCR or western hybridization methods. Gross phenotypic changes are not the norm, unless the recombinant gene has been chosen and modified so as to generate biological activity in the host organism. Additional phenotypes that are encountered include toxicity to the host organism induced by the recombinant gene product, especially if it is over-expressed or expressed within inappropriate cells or tissues.

In some cases, recombinant DNA can have deleterious effects even if it is not expressed. One mechanism by which this happens is insertional inactivation, in which the rDNA becomes inserted into a host cell's gene. In some cases, researchers use this phenomenon to "knock out" genes to determine their biological function and importance. Another mechanism by which rDNA insertion into chromosomal DNA can affect gene expression is by inappropriate activation of previously unexpressed host cell genes. This can happen, for example, when a recombinant DNA fragment containing an active promoter becomes located next to a previously silent host cell gene, or when a host cell gene that functions to restrain gene expression undergoes insertional inactivation by recombinant DNA.

Applications of recombinant DNA

Recombinant DNA is widely used in biotechnology, medicine and research. Today, recombinant proteins and other products that result from the use of DNA technology are found in essentially every western pharmacy, physician or veterinarian office, medical testing laboratory, and biological research laboratory. In addition, organisms that have been manipulated using recombinant DNA technology, as well as products derived from those organisms, have found their way into many farms, supermarkets, home medicine cabinets, and even pet shops, such as those that sell GloFish and other genetically modified animals.

The most common application of recombinant DNA is in basic research, in which the technology is important to most current work in the biological and biomedical sciences. Recombinant DNA is used to identify, map and sequence genes, and to determine their function. rDNA probes are employed in analyzing gene expression within individual cells, and throughout the tissues of whole organisms. Recombinant proteins are widely used as reagents in laboratory experiments and to generate antibody probes for examining protein synthesis within cells and organisms.

Many additional practical applications of recombinant DNA are found in industry, food production, human and veterinary medicine, agriculture, and bioengineering. Some specific examples are identified below.

Recombinant chymosin

Found in rennet, chymosin is an enzyme required to manufacture cheese. It was the first genetically engineered food additive used commercially. Traditionally, processors obtained chymosin from rennet, a preparation derived from the fourth stomach of milk-fed calves. Scientists engineered a non-pathogenic strain (K-12) of *E. coli* bacteria for large-scale laboratory production of the enzyme. This microbiologically produced recombinant enzyme, identical structurally to the calf derived enzyme, costs less and is produced in abundant quantities. Today about 60% of U.S. hard cheese is made with genetically engineered chymosin. In 1990, FDA granted chymosin "generally recognized as safe" (GRAS) status based on data showing that the enzyme was safe.

Recombinant human insulin

Almost completely replaced insulin obtained from animal sources (e.g. pigs and cattle) for the treatment of insulin-dependent diabetes. A variety of different recombinant insulin preparations are in widespread use. Recombinant insulin is synthesized by inserting the human insulin gene into *E. coli*, or yeast (Saccharomyces cerevisiae) which then produces insulin for human use. Insulin produced by E. coli requires further post translational modifications (glycosylation) whereas yeast are able to perform these themselves by virtue of being a more complex host organism. The advantage of recombinant human insulin is after chronic use patients don't develop an immune defence against it the way animal sourced insulin stimulates the human immune system.

Recombinant human growth hormone (HGH, somatotropin)

Administered to patients whose pituitary glands generate insufficient quantities to support normal growth and development. Before recombinant HGH became available, HGH for therapeutic use was obtained from pituitary glands of cadavers. This unsafe practice led to some patients developing Creutzfeldt–Jakob disease. Recombinant HGH eliminated this problem, and is now used therapeutically. It has also been misused as a performance-enhancing drug by athletes and others. DrugBank entry

Recombinant blood clotting factor VIII

A blood-clotting protein that is administered to patients with forms of the bleeding disorder hemophilia, who are unable to produce factor VIII in quantities sufficient to support normal blood coagulation. Before the development of recombinant factor VIII, the protein was obtained by processing large quantities of human blood from multiple donors, which carried a very high risk of transmission of blood borne infectious diseases, for example HIV and hepatitis B. DrugBank entry

Recombinant hepatitis B vaccine

Hepatitis B infection is controlled through the use of a recombinant hepatitis B vaccine, which contains a form of the hepatitis B virus surface antigen that is produced in yeast cells. The development of the recombinant subunit vaccine was an important and necessary development because hepatitis B virus, unlike other common viruses such as polio virus, cannot be grown in vitro. Vaccine information from Hepatitis B Foundation

Recombinant antibodies

Recombinant antibodies (rAbs) are produced in vitro by the means of expression systems based on mammalian cells. Their monospecific binding to a specific epitope makes rAbs eligible not only for research purposes, but also as therapy options against certain cancer types, infections and autoimmune diseases.

Diagnosis of infection with HIV

Each of the three widely used methods for diagnosing HIV infection has been developed using recombinant DNA. The antibody test (ELISA or western blot) uses a recombinant HIV protein to test for the presence of antibodies that the body has produced in response to an HIV infection. The DNA test looks for the presence of HIV genetic material using reverse transcription polymerase chain reaction (RT-PCR). Development of the RT-PCR test was made possible by the molecular cloning and sequence analysis of HIV genomes. HIV testing page from US Centers for Disease Control (CDC)

Golden rice

A recombinant variety of rice that has been engineered to express the enzymes responsible for â-carotene biosynthesis. This variety of rice holds substantial promise for reducing the incidence of vitamin A deficiency in the world's population. Golden rice is not currently in use, pending the resolution of regulatory and intellectual property issues.

Herbicide-resistant crops

Commercial varieties of important agricultural crops (including soy, maize/corn, sorghum, canola, alfalfa and cotton) have been developed that incorporate a recombinant gene that results in resistance to the herbicide glyphosate (trade name *Roundup*), and simplifies weed control by glyphosate application. These crops are in common commercial use in several countries.

Insect-resistant crops

Bacillus thuringeiensis is a bacterium that naturally produces a protein (Bt toxin) with insecticidal properties. The bacterium has been applied to crops as an insect-control strategy for many years, and this practice has been widely adopted in agriculture and gardening. Recently, plants have been developed that express a recombinant form of the bacterial protein, which may effectively control some insect predators. Environmental issues associated with the use of these transgenic crops have not been fully resolved.

History

The idea of recombinant DNA was first proposed by Peter Lobban, a graduate student of Prof. Dale Kaiser in the Biochemistry Department at Stanford University Medical School. The first publications describing the successful production and intracellular replication of recombinant DNA appeared in 1972 and 1973, from Stanford and UCSF. In 1980 Paul Berg, a professor in the Biochemistry Department at Stanford and an author on one of the first papers was awarded the Nobel Prize in Chemistry for his work on nucleic acids "with particular regard to recombinant DNA". Werner Arber, Hamilton Smith, and Daniel Nathans shared the 1978 Nobel Prize in Physiology or Medicine for the discovery of restriction endonucleases which enhanced the techniques of rDNA technology.

Stanford University applied for a US patent on recombinant DNA in 1974, listing the inventors as Herbert W. Boyer (professor at the University of California, San Francisco) and Stanley N. Cohen (professor at Stanford University); this patent was awarded in 1980. The first licensed drug generated using recombinant DNA technology was human insulin, developed by Genentech and licensed by Eli Lilly and Company.

Controversy

Scientists associated with the initial development of recombinant DNA methods recognized that the potential existed for organisms containing recombinant DNA to have undesirable or dangerous properties. At the 1975 Asilomar Conference on Recombinant DNA, these concerns were discussed and a voluntary moratorium on recombinant DNA research was initiated for experiments that were considered particularly risky. This moratorium was widely observed until the National Institutes of Health (USA) developed and issued formal guidelines for rDNA work. Today, recombinant DNA molecules and recombinant proteins are usually not regarded as dangerous. However, concerns remain about some organisms that express recombinant DNA, particularly when they leave the laboratory and are introduced into the environment or food chain. These concerns are discussed in the articles on genetically modified organisms and genetically modified food controversies. Furthermore, there are concerns about the by-products in biopharmaceutical production, where recombinant DNA result in specific protein products. The major by-product, termed host cell protein, comes from the host expression system and poses a threat to the patient's health and the overall environment.

Development Genetics

The branch of genetics that studies how genes control embryonic development. Advances in the field of developmental genetics have emphasized the degree of conservation throughout evolution of the genes (basic units in inheritance consisting of DNA) that control development

Developmental biology is the study of the process by which animals and plants grow and develop. Developmental biology also encompasses the biology of regeneration, asexual reproduction, metamorphosis, and the growth and differentiation of stem cells in the adult organism.

Perspectives

The main processes involved in the embryonic development of animals are: tissue patterning (via regional specification and patterned cell differentiation); tissue growth; and tissue morphogenesis.

- Regional specification refers to the processes that create the spatial patterns in a ball or sheet of initially similar cells. This generally involves the action of cytoplasmic determinants, located within parts of the fertilized egg, and of inductive signals emitted from signaling centers in the embryo. The early stages of regional specification do not generate functional differentiated cells, but cell populations committed to developing to a specific region or part of the organism. These are defined by the expression of specific combinations of transcription factors.
- Cell differentiation relates specifically to the formation of functional cell types such as nerve, muscle, secretory epithelia, etc. Differentiated cells contain large amounts of specific proteins associated with cell function.
- Morphogenesis relates to the formation of a three-dimensional shape. It mainly involves the orchestrated movements of cell sheets and of individual cells. Morphogenesis is important for creating the three germ layers of the early embryo (ectoderm, mesoderm, and endoderm) and for building up complex structures during organ development.
- Tissue growth involves both an overall increase in tissue size, and also the differential growth of parts (allometry) which contributes to morphogenesis. Growth mostly occurs through cell proliferation but also through changes in cell size or the deposition of extracellular materials.

The development of plants involves similar processes to that of animals. However, plant cells are mostly immotile so morphogenesis is achieved by differential growth, without cell movements. Also, the inductive signals and the genes involved are different from those that control animal development.

Developmental processes

Cell differentiation

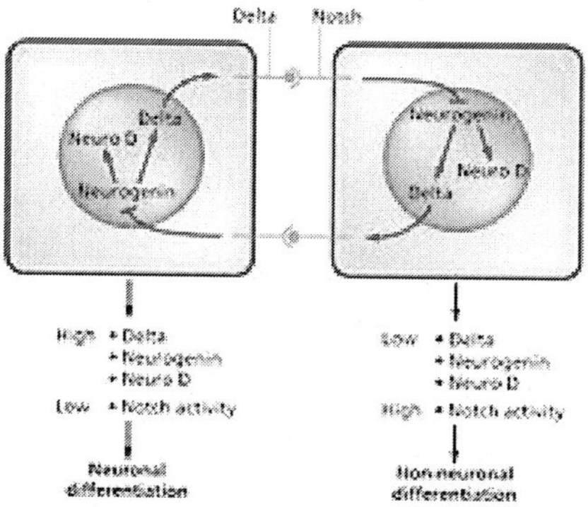

The Notch-delta system in neurogenesis.

Cell differentiation is the process whereby different functional cell types arise in development. For example, neurons, muscle fibers and hepatocytes (liver cells) are well known types of differentiated cells. Differentiated cells usually produce large amounts of a few proteins that are required for their specific function and this gives them the characteristic appearance that enables them to be recognized under the light microscope. The genes encoding these proteins are highly active. Typically their chromatin structure is very open, allowing access for the transcription enzymes, and specific transcription factors bind to regulatory sequences in the DNA in order to activate gene expression. For example, NeuroD is a key transcription factor for neuronal differentiation, myogenin for muscle differentiation, and HNF4 for hepatocyte differentiation. Cell differentiation is usually the final stage of development, preceded by several states of commitment which are not visibly differentiated. A single tissue, formed from a single type of progenitor cell or stem cell, often consists of several differentiated cell types. Control of their formation involves a process of lateral inhibition, based on the properties of the Notch signaling pathway. For example, in the neural plate of the embryo this system operates to generate a population of neuronal precursor cells in which NeuroD is highly expressed.

Regeneration

Regeneration indicates the ability to regrow a missing part. This is very prevalent amongst plants, which show continuous growth, and also among colonial animals such as hydroids and ascidians. But most interest by developmental biologists has been shown in the regeneration of parts in free living animals. In particular four models have been the subject of much investigation. Two of these have the ability to regenerate whole bodies: *Hydra*, which can regenerate any part of the polyp from a small fragment, and planarian worms, which can usually regenerate both heads and tails. Both of these examples have continuous cell turnover fed by stem cells and, at least in planaria, at least some of the stem cells have been shown to be pluripotent. The other two models show only distal regeneration of appendages. These are the insect appendages, usually the legs of hemimetabolous insects such as the cricket, and the limbs of urodele amphibians. Considerable information is now available about amphibian limb regeneration and it is known that each cell type regenerates itself, except for connective tissues where there is considerable interconversion between cartilage, dermis and tendons. In terms of the pattern of structures, this is controlled by a re-activation of signals active in the embryo. There is still debate about the old question of whether regeneration is a "pristine" or an "adaptive" property. If the former is the case, with improved knowledge, we might expect to be able to improve regenerative ability in humans. If the latter, then each instance of regeneration is presumed to have arisen by natural selection in circumstances particular to the species, so no general rules would be expected.

Embryonic development of animals

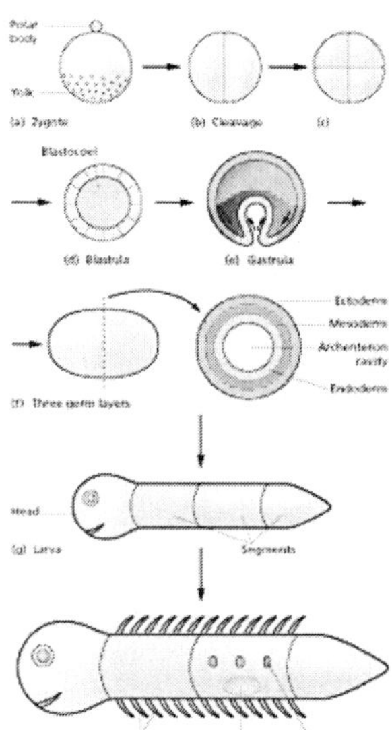

Generalized scheme of embryonic development. Slack "Essential Developmental Biology"

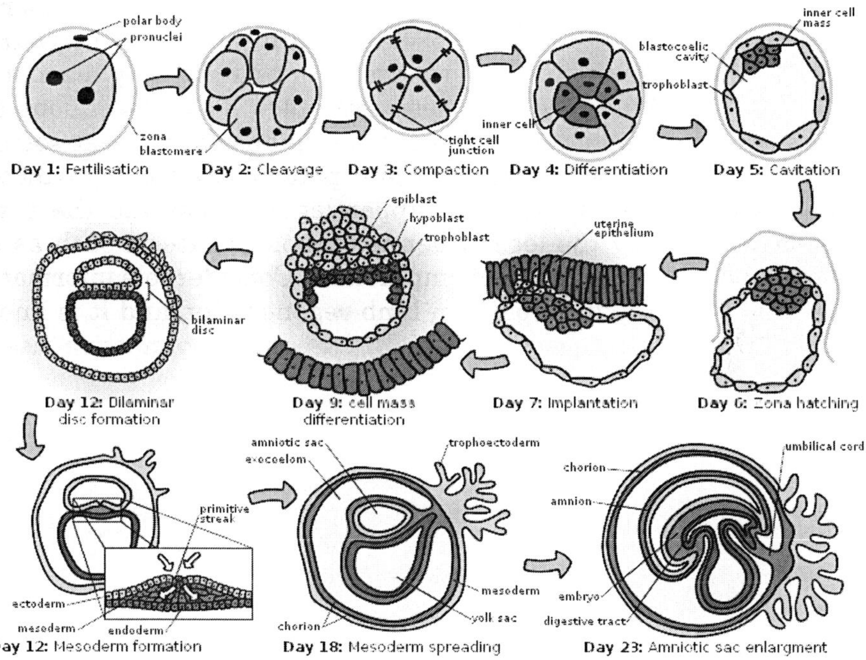

The initial stages of human embryogenesis.

The sperm and egg fuse in the process of fertilization to form a fertilized egg, or zygote. This undergoes a period of divisions to form a ball or sheet of similar cells called a blastula or blastoderm. These cell divisions are usually rapid with no growth so the daughter cells are half the size of the mother cell and the whole embryo stays about the same size. They are called cleavage divisions.

Mouse epiblast primordial germ cells undergo extensive epigenetic reprogramming. This process involves genome-wide DNA demethylation, chromatin reorganization and epigenetic imprint erasure leading to totipotency. DNA demethylation is carried out by a process that utilizes the DNA base excision repair pathway.

Morphogenetic movements convert the cell mass into a three layered structure consisting of multicellular sheets called ectoderm, mesoderm and endoderm. These sheets are known as germ layers. This is the process of gastrulation. During cleavage and gastrulation the first regional specification events occur. In addition to the formation of the three germ layers themselves, these often generate extraembryonic structures, such as the mammalian placenta, needed for support and nutrition of the embryo, and also establish differences of commitment along the anteroposterior axis (head, trunk and tail).

Regional specification is initiated by the presence of cytoplasmic determinants in one part of the zygote. The cells that contain the determinant become a signaling center and emit an inducing factor. Because the inducing factor is produced in one place, diffuses away, and decays, it forms a concentration gradient, high near the source cells and low further away. The remaining cells of the embryo, which do not contain the determinant, are competent to respond to different concentrations by upregulating specific developmental control genes. This results in a series of zones becoming set up, arranged at progressively greater distance from the signaling center. In each zone a different combination of developmental control genes is upregulated. These genes encode transcription factors which upregulate new combinations of gene activity in each region. Among other functions, these transcription factors control expression of genes conferring specific adhesive and motility properties on the cells in which they are active. Because of these different morphogenetic properties, the cells of each germ layer move to form sheets such that the ectoderm ends up on the outside, mesoderm in the middle, and endoderm on the inside. Morphogenetic movements not only change the shape and structure of the embryo, but by bringing cell sheets into new spatial relationships they also make possible new phases of signaling and response between them.

Growth in embryos is mostly autonomous. For each territory of cells the growth rate is controlled by the combination of genes that are active. Free-living embryos do not grow in mass as they have no external food supply. But embryos fed by a placenta or extraembryonic yolk supply can grow very fast, and changes to relative growth rate between parts in these organisms help to produce the final overall anatomy.

The whole process needs to be coordinated in time and how this is controlled is not understood. There may be a master clock able to communicate with all parts of the embryo that controls the course of events, or timing may depend simply on local causal sequences of events.

Metamorphosis

Developmental processes are very evident during the process of metamorphosis. This occurs in various types of animal. Well-known examples are seen in frogs, which usually hatch as a tadpole and metamorphoses to an adult frog, and certain insects which hatch as a larva and then become remodeled to the adult form during a pupal stage.

All the developmental processes listed above occur during metamorphosis. Examples that have been especially well studied include tail loss and other changes in the tadpole of the frog *Xenopus*, and the biology of the imaginal discs, which generate the adult body parts of the fly *Drosophila melanogaster*.

Genomics & Proteomics - Genetics on the internet

Human cardiovascular disease is a consequence of a complex interplay of genetic, epigenetic and environmental factors. The resulting phenotype has traditionally been described by a variety of clinical descriptors, as well as biochemical markers and imaging techniques. To date, conventional cardiology attempts to prevent disease by modifying established risk factors and to treat manifest disease, often after complications have occurred and irreversible tissue changes have taken place. Current therapies (eg, those for heart failure) are administered uniformly across a heterogeneous spectrum of disease etiology, severity and genetic background. As large-scale investigation platforms become more mainstream, we have the opportunity to expand our medical armamentarium to include molecular phenotypes that distinguish subtle sub-classifications of disease and allow us to better tailor both prevention strategies and therapeutics. The present paper reviews the potential contributions of genomic and proteomic approaches in developing a more personalized approach to cardiovascular medicine.

PANOMIC APPROACHES TO BIOLOGY

The emergence of high-throughput molecular technologies has spurred a paradigm shift from a reductionist, single-molecule, biological approach to one that integrates vast networks of information into fluid, dynamic systems. 'Omics' encompasses comprehensive methodologies that attempt to capture the exhaustive output of an organism's genes (genomics), RNA (transcriptomics), metabolites (metabolomics) and proteins (proteomics). The whole is often greater than the sum of its parts, and by embracing the complexity of various profiling patterns, far more subtle information may be extracted than would have been otherwise available from single targets. The systems biology approach allows the study of networks of interactions, in addition to dissecting individual molecular components. This approach has a long history, but has only recently received renewed attention given the increasing availability of pathway data and advances in computational analyses.

Complementary to this global, nonhypothesis-driven methodology is the candidate gene approach, which evaluates genes and proteins based on a prior understanding of their roles in disease. Understanding a system's structure and dynamics has proven to be a multidisciplinary venture, requiring expertise in the fields of biology, engineering, chemistry and informatics. Its ultimate impact on cardiovascular medicine remains to be determined, but will likely be far-reaching, leading to innovations in many aspects of patient care. The present paper introduces the reader to the fundamentals of proteomics, and refers the reader to comprehensive reviews detailing genomic and transcriptomic methodologies.

PROTEOMICS: GENERAL CONSIDERATIONS

Proteomics is the study of the proteome, or the entire protein complement of a genome. The proteome consists of all proteins present in a cell or tissue at a given time and is far more complex than was originally proposed by the one-gene, one-transcript, one-protein hypothesis. On average, there are five to seven protein isoforms per open reading frame in the human genome. To date, it is apparent that the approximately 30,000 human genes encode for nearly one million proteins. Alternative splicing is a significant contributor to protein diversity, occurring in 35% to 60% of our genes. This is complemented by other mechanisms, such as the use of multiple transcription start sites, polyadenylation and specific editing of pre-messenger RNA. The diversity is further compounded by the post-translational modification of proteins, such as phosphorylation, sulphation, glycosylation, hydroxylation, N-methylation, carboxymethylation, acetylation, prenylation and N-myristylation.

Proteins are the effectors of nearly every cellular function and, as such, dictate the phenotype of a given cell, tissue or organ, representing not only the genetic composition of an individual, but gene-gene and gene-environment interactions. The proteome varies under all physiological conditions, with pronounced changes observed in aging, as well as acute or chronic stressors leading to disease states. Chronic illnesses often result in altered protein levels due to specific gene up- or downregulation, isoform switching or de novo protein synthesis. There is often insufficient time during an acute illness to perturb the synthetic machinery of proteins, and the most common mechanisms of protein alteration are, therefore, post-translational modifications.

Numerous techniques exist for the analysis of the proteome. The development of two-dimensional polyacrylamide gel electrophoresis (2-DE) more than 20 years ago permitted the simultaneous separation of large numbers of proteins first by charge (isoelectric point) and then by molecular mass. Difficulties arise with 2-DE when separating low-abundance or membrane-associated proteins, or those at the extremes of pH or molecular weight. A complementary approach uses isotope-coded affinity tags, which selectively label cysteine residues of peptide fragments following tryptic digest of the protein sample. This increases the sensitivity of the detection of smaller peptides and permits quantification of changes in protein composition. Gel-free methods include 'shotgun' approaches using tryptic digests of complex sets of proteins, followed by peptide sequencing in a tandem mass spectrometer for identification . More recently, state-of-the-art liquid chromatography-mass spectroscopy systems have been used to identify and track the relative abundance of thousands of proteins. Despite the rapid advances in proteomic technology, several areas continue to need improvement, particularly in the detection, quantification and identification of low-abundance proteins and assessment of protein distribution among cells and subcellular structures, as well as the characterization of post-translational modifications.

Biological samples subjected to proteomic analysis comprise three major types: tissues, cell populations and biological fluids. To date, most of the cardiovascular studies have focused on tissue obtained by endomyocardial biopsy or, more commonly, cardiac explantation at the time of transplantation. Endomyocardial biopsy carries a small but significant degree of risk inherent in the procedure. As well, the small sample volume is at the lower limit of detection for most high-throughput technologies today. It is, therefore, important to consider the utility of molecular signature analysis derived from peripheral blood leukocytes. Successful examples of this exist in the cancer literature, and more recently, peripheral blood molecular signatures correlated with biopsy-proven allograft rejection in cardiac transplant recipients . The dynamic range of proteins in the blood is formidable, ranging from 55,000,000,000 pg/mL (eg, albumin) to 5 pg/mL (eg, interleukin-6). Only four proteins (albumin, transferrin, haptoglobulin and immunoglobulins) make up more than 90% of the protein mass in the blood . Although techniques exist for depleting these proteins from the serum, challenges remain to assure their efficiency and specificity.

Essential aids to proteomic studies of the heart are the federated online databases of human cardiac proteins that have been established. These databases, known as HSC-2DPAGE , HEART-2DPAGE and HP-2DPAGE , are widely accessible and display more than 6000 myocardial proteins separated by 2-DE. Open-access data sharing through such mechanisms will be crucial for the rapid advancement of the field and international collaboration. There are also international organizations attempting to provide standards and benchmarks for the increasing data being mined in proteomic studies. One organization, the Human Proteome Organisation, was created in 2001 with the objective of going beyond the Human Genome Project in understanding health and disease, while providing clear guidelines for proteomic studies . A second key organization is the Proteomics Division at the National Heart, Lung and Blood Institute .

Proteomics

Proteins are the final products of genes that perform the function encoded by the gene. Proteins are composed of amino acids and play important roles in the cell. All enzymes (except ribozymes) are proteins and act as catalysts that affect the rate of reactions. Proteins are also regulatory molecules, and some are hormones. Transport proteins, such as hemoglobin, help transport oxygen to various organs. Antibodies that defend against foreign particles are also proteins. In the diseased state, protein function can be impaired because of changes at the genetic level or because of direct impact on a specific protein.

A proteome is the entire set of proteins produced by a cell type. Proteomes can be studied using the knowledge of genomes because genes code for mRNAs, and the mRNAs encode proteins. The study of the function of proteomes is called proteomics. Proteomics complements genomics and is useful when scientists want to test their hypotheses that were based on genes. Even though all cells in a multicellular organism have the same set of genes, the set of proteins produced in different tissues is different and dependent on gene expression. Thus, the genome is constant, but the proteome varies and is dynamic within an organism. In addition, RNAs can be alternatively spliced (cut and pasted to create novel combinations and novel proteins), and many proteins are modified after translation. Although the genome provides a blueprint, the final architecture depends on several factors that can change the progression of events that generate the proteome.

Genomes and proteomes of patients suffering from specific diseases are being studied to understand the genetic basis of the disease. The most prominent disease being studied with proteomic approaches is cancer. Proteomic approaches are being used to improve the screening and early detection of cancer; this is achieved by identifying proteins whose expression is affected by the disease process. An individual protein is called a biomarker, whereas a set of proteins with altered expression levels is called a protein signature. For a biomarker or protein signature to be useful as a candidate for early screening and detection of a cancer, it must be secreted in body fluids such as sweat, blood, or urine, so that large-scale screenings can be performed in a noninvasive fashion. The current problem with using biomarkers for the early detection of cancer is the high rate of false-negative results. A false-negative result is a negative test result that should have been positive. In other words, many cases of cancer go undetected, which makes biomarkers unreliable. Some examples of protein biomarkers used in cancer detection are CA-125 for ovarian cancer and PSA for prostate cancer. Protein signatures may be more reliable than biomarkers to detect cancer cells. Proteomics is also being used to develop individualized treatment plans, which involves the prediction of whether or not an individual will respond to specific drugs and the side effects that the individual may have. Proteomics is also being used to predict the possibility of disease recurrence.

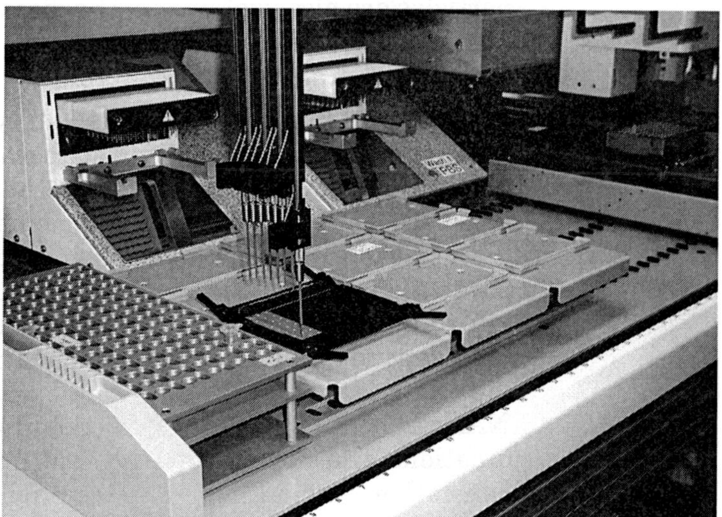

Figure : This machine is preparing to do a proteomic pattern analysis to identify specific cancers so that an accurate cancer prognosis can be made.

The National Cancer Institute has developed programs to improve the detection and treatment of cancer. The Clinical Proteomic Technologies for Cancer and the Early Detection Research Network are efforts to identify protein signatures specific to different types of cancers. The Biomedical Proteomics Program is designed to identify protein signatures and design effective therapies for cancer patients.

Chapter 2

Mendelism : Mendel's Law of Inheritance

Introduction to Mendelism

Mendelism:

1. Mendelism is a term that refers to the single-gene trait's theoretical principles of heredity, which are taken from Gregor Mendel's principles, also known as Mendel's laws.
2. Gregor Mendel conducted several experiments on pea plants for seven years and then postulated the laws of inheritance.
3. Mendel's law of inheritance includes the law of dominance, the law of segregation, and the law of independent assortment.

Law of dominance:

1. The first law of inheritance is the law of dominance.
2. The law states that hybrid offspring will only inherit the dominant characteristics in the phenotype.
3. The alleles that suppress a trait are recessive traits, whereas the alleles that define a trait are known as dominant traits.
4. Monohybrid cross helped in proposing the law of dominance.
5. In a monohybrid cross, Mendel crossed two pure pea plants with opposing traits (one short and one tall).
6. Mendel named those first-generation offsprings as F_1 progeny and all the offspring produced by crossing tall plants (dominant trait) and dwarf plants (recessive trait) were tall (heterozygous dominant or hybrid).
7. After that, Mendel crossed F_1 progeny, and both tall and dwarf plants were obtained in the F_2 generation.

Law of segregation:

1. The law states that every pair of alleles or genes in parents, divides and a single gene is transferred from each parent to their offspring. The passage of a particular gene in a pair is a chance event.
2. There are several alleles for a single gene.
3. Every organism inherits a pair of alleles for each gene.
4. During the process of gametogenesis, when gametes are produced by meiosis, Then the allelic pairs segregate, leaving each gamete with only one allele.
5. A gene has two alternate forms known as alleles. One allele is dominant while the other allele is recessive.

Law of independent assortment:

1. According to this law, discrete pairs of alleles pass onto the children without being dependent on one another and as a result, the inheritance of genes in a particular region of the genome has no effect on the inheritance of genes in another region.
2. A dihybrid cross was used to formulate the law of independent assortment.
3. In a dihybrid experiment, Mendel selected two traits, each with two alleles.
4. He crossed round-yellow seeds (dominant) and wrinkled-green seeds (recessive) and observed that all of the F_1 progeny were round-yellow.
5. He then self-pollinated the F_1 progeny and obtained four distinct phenotypes that are round-yellow, round-green, wrinkled-yellow, and wrinkled-green seeds in the phenotypic ratio of 9:3:3:1.

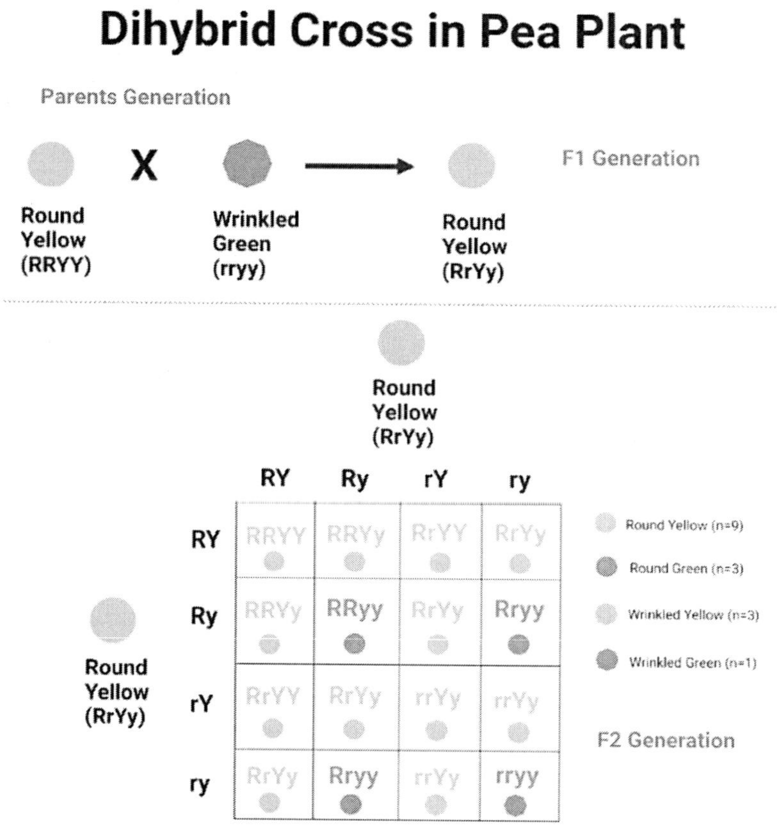

Mendel's Laws of Inheritance

Inheritance can be defined as the process of how a child receives genetic information from the parent. The whole process of heredity is dependent upon inheritance and it is the reason that the offsprings are similar to the parents. This simply means that due to inheritance, the members of the same family possess similar characteristics.

It was only during the mid 19th century that people started to understand inheritance in a proper way. This understanding of inheritance was made possible by a scientist named Gregor Mendel, who formulated certain laws to understand inheritance known as Mendel's laws of inheritance.

Mendel's Laws of Inheritance

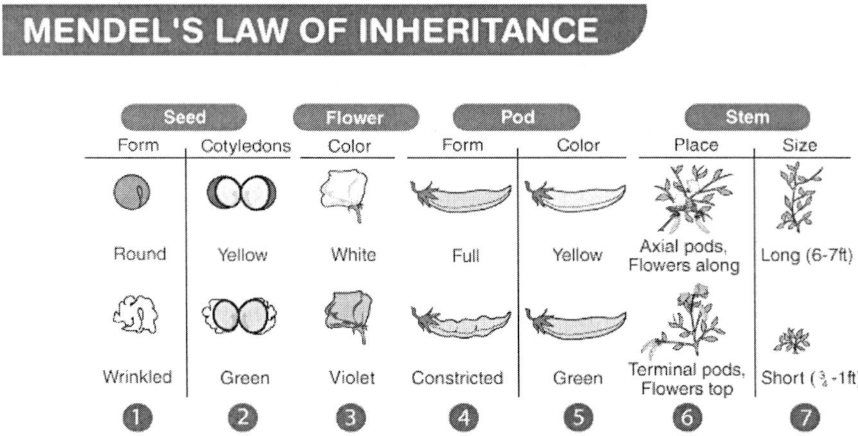

Between 1856-1863, Mendel conducted the hybridization experiments on the garden peas. During that period, he chose some distinct characteristics of the peas and conducted some cross-pollination/artificial pollination on the pea lines that showed stable trait inheritance and underwent continuous self-pollination. Such pea lines are called true-breeding pea lines.

Why was Pea Plant Selected for Mendel's Experiments?

He selected a pea plant for his experiments for the following reasons:

1. The pea plant can be easily grown and maintained.
2. They are naturally self-pollinating but can also be cross-pollinated.
3. It is an annual plant, therefore, many generations can be studied within a short period of time.
4. It has several contrasting characters.

Mendel conducted 2 main experiments to determine the laws of inheritance. These experiments were:

1. Monohybrid Cross
2. Dihybrid Cross

While experimenting, Mendel found that certain factors were always being transferred down to the offspring in a stable way. Those factors are now called genes i.e. genes can be called the units of inheritance.

Mendel's Experiments

Mendel experimented on a pea plant and considered 7 main contrasting traits in the plants. Then, he conducted both experiments to determine the inheritance laws. A brief explanation of the two experiments is given below.

Monohybrid Cross

In this experiment, Mendel took two pea plants of opposite traits (one short and one tall) and crossed them. He found the first generation offspring were tall and called it F1 progeny. Then he crossed F1 progeny and obtained both tall and short plants in the ratio 3:1. To know more about this experiment, visit Monohybrid Cross – Inheritance Of One Gene.

Mendel even conducted this experiment with other contrasting traits like green peas vs yellow peas, round vs wrinkled, etc. In all the cases, he found that the results were similar. From this, he formulated the **laws of Segregation And Dominance.**

Dihybrid Cross

In a dihybrid cross experiment, Mendel considered two traits, each having two alleles. He crossed wrinkled-green seed and round-yellow seeds and observed that all the first generation progeny (F1 progeny) were round-yellow. This meant that dominant traits were the round shape and yellow colour.

He then self-pollinated the F1 progeny and obtained 4 different traits: round-yellow, round-green, wrinkled-yellow, and wrinkled-green seeds in the ratio 9:3:3:1.

GREGOR MENDEL'S LIFE (1822-1884)

Gregor Mendel, in full **Gregor Johann Mendel**, original name (until 1843) **Johann Mendel**, (born July 20, 1822, Heinzendorf, Silesia, Austrian Empire [now Hynèice, Czech Republic]— died January 6, 1884, Brünn, Austria-Hungary [now Brno, Czech Republic]), botanist, teacher, and Augustinian prelate, the first person to lay the mathematical foundation of the science of genetics, in what came to be called Mendelism.

Education and early career

Born to a family with limited means in German-speaking Silesia, Mendel was raised in a rural setting. His academic abilities were recognized by the local priest, who persuaded his parents to send him away to school at the age of 11. His *Gymnasium* (grammar school) studies completed in 1840, Mendel entered a two-year program in philosophy at the Philosophical Institute of the University of Olmütz (Olomouc, Czech Republic), where he excelled in physics and mathematics, completing his studies in 1843. His initial years away from home were hard, because his family could not sufficiently support him. He tutored other students to make ends meet, and twice he suffered serious depression and had to return home to recover. As his father's only son, Mendel was expected to take over the small family farm, but he preferred a different solution to his predicament, choosing to enter the Altbrünn monastery as a novitiate of the Augustinian order, where he was given the name Gregor.

The move to the monastery took him to Brünn, the capital of Moravia, where for the first time he was freed from the harsh struggle of former years. He was also introduced to a diverse and intellectual community. As a priest, Mendel found his parish duty to visit the sick and dying so distressing that he again became ill. Abbot Cyril Napp found him a substitute-teaching position at Znaim (Znojmo, Czech Republic), where he proved very successful. However, in 1850 Mendel failed an exam—introduced through new legislation for teacher certification—and was sent to the University of Vienna for two years to benefit from a new program of scientific instruction. As at Olmütz, Mendel devoted his time at Vienna to physics and mathematics, working under Austrian physicist Christian Doppler and mathematical physicist Andreas von Ettinghausen. He also studied the anatomy and physiology of plants and the use of the microscope under botanist Franz Unger, an enthusiast for the cell theory and a supporter of the developmentalist (pre-Darwinian) view of the evolution of life. Unger's writings on the latter made him a target for attack by the Roman Catholic press of Vienna shortly before and during Mendel's time there.

In the summer of 1853, Mendel returned to the monastery in Brünn, and in the following year he was again given a teaching position, this time at the Brünn Realschule (secondary school), where he remained until elected abbot 14 years later. He attempted the teacher exam again in 1856, although the event caused a nervous breakdown and a second failure. However, these years were his greatest in terms of success both as teacher and as consummate experimentalist. Once abbot, his administrative duties came to occupy the majority of his time. Moreover, Mendel's refusal to permit the monastery to pay the state's new taxes for a religious fund led to his involvement in a long and bitter dispute with the authorities. Convinced that this tax was unconstitutional, he continued his opposition, refusing to comply even when the state took over the administration of some of the monastery's estates and directed the profits to the religious fund.

Experimental period

In 1854 Abbot Cyril Napp permitted Mendel to plan a major experimental program in hybridization at the monastery. The aim of this program was to trace the transmission of hereditary characters in successive generations of hybrid progeny. Previous authorities had observed that progeny of fertile hybrids tended to revert to the originating species, and they had therefore concluded that hybridization could not be a mechanism used by nature to multiply species—though in exceptional cases some fertile hybrids did appear not to revert (the so-called "constant hybrids"). On the other hand, plant and animal breeders had long shown that crossbreeding could indeed produce a multitude of new forms. The latter point was of particular interest to landowners, including the abbot of the monastery, who was concerned about the monastery's future profits from the wool of its Merino sheep, owing to competing wool being supplied from Australia.

Mendel chose to conduct his studies with the edible pea (Pisum sativum) because of the numerous distinct varieties, the ease of culture and control of pollination, and the high proportion of successful seed germinations. From 1854 to 1856 he tested 34 varieties for constancy of their traits. In order to trace the transmission of characters, he chose seven traits that were expressed in a distinctive manner, such as plant height (short or tall) and seed colour (green or yellow). He referred to these alternatives as contrasted characters, or character-pairs. He crossed varieties that differed in one trait—for instance, tall crossed with short. The first generation of hybrids (F_1) displayed the character of one variety but not that of the other. In Mendel's terms, one character was dominant and the other recessive. In the numerous progeny that he raised from these hybrids (the second generation, F_2), however, the recessive character reappeared, and the proportion of offspring bearing the dominant to offspring bearing the recessive was very close to a 3 to 1 ratio. Study of the descendants (F_3) of the dominant group showed that one-third of them were true-breeding and two-thirds were of hybrid constitution. The 3:1 ratio could hence be rewritten as 1:2:1, meaning that 50 percent of the F_2 generation were true-breeding and 50 percent were still hybrid. This was Mendel's major discovery, and it was unlikely to have been made by his predecessors, since they did not grow statistically significant populations, nor did they follow the individual characters separately to establish their statistical relations.

Mendel's approach to experimentation came from his training in physics and mathematics, especially combinatorial mathematics. The latter served him ideally to represent his result. If A represents the dominant characteristic and a the recessive, then the 1:2:1 ratio recalls the terms in the expansion of the binomial equation:$(A + a)^2 = A^2 + 2Aa + a^2$Mendel realized further that he could test his expectation that the seven traits are transmitted independently of one another. Crosses involving first two and then three of his seven traits yielded categories of offspring in proportions following the terms produced from combining two binomial equations, indicating that their transmission was independent of one another. Mendel's successors have called this conclusion the law of independent assortment.

Theoretical interpretation

Mendel went on to relate his results to the cell theory of fertilization, according to which a new organism is generated from the fusion of two cells. In order for pure breeding forms of both the dominant and the recessive type to be brought into the hybrid, there had to be some temporary accommodation of the two differing characters in the hybrid as well as a separation process in the formation of the pollen cells and the egg cells. In other words, the hybrid must form germ cells bearing the potential to yield either the one characteristic or the other. This has since been described as the law of segregation, or the doctrine of the purity of the germ cells. Since one pollen cell fuses with one egg cell, all possible combinations of the differing pollen and egg cells would yield just the results suggested by Mendel's combinatorial theory.

Mendel first presented his results in two separate lectures in 1865 to the Natural Science Society in Brünn. His paper "Experiments on Plant Hybrids" was published in the society's journal, *Verhandlungen des naturforschenden Vereines in Brünn*, the following year. It attracted little attention, although many libraries received it and reprints were sent out. The tendency of those who read it was to conclude that Mendel had simply demonstrated more accurately what was already widely assumed—namely, that hybrid progeny revert to their originating forms. They overlooked the potential for variability and the evolutionary implications that his demonstration of the recombination of traits made possible. Most notably, Swiss botanist Karl Wilhelm von Nägeli actually corresponded with Mendel, despite remaining skeptical as to the significance of his results and doubting that the germ cells in hybrids could be pure.

Latter years

Mendel appears to have made no effort to publicize his work, and it is not known how many reprints of his paper he distributed. He had ordered 40 reprints, the whereabouts of only eight of which are known. Other than the journal that published his paper, 15 sources are known from the 19th century in which Mendel is mentioned in the context of plant hybridization. Few of these provide a clear picture of his achievement, and most are very brief.

By 1871 Mendel had only enough time to continue his meteorological and apicultural work. He traveled little, and his only visit to England was to see the Industrial Exhibition in 1862. Bright disease made his last years painful, and he died at the age of 61. Mendel's funeral was attended by many mourners and proceeded from the monastery to the monastery's burial plot in the town's central cemetery, where his grave can be seen today. He was survived by two sisters and three nephews.

Rediscovery

In 1900 Dutch botanist and geneticist Hugo de Vries, German botanist and geneticist Carl Erich Correns, and Austrian botanist Erich Tschermak von Seysenegg independently reported results of hybridization experiments similar to Mendel's, though each later claimed not to have known of Mendel's work while doing their own experiments. However, both de Vries and Correns had read Mendel earlier—Correns even made detailed notes on the subject—but had forgotten. De Vries had a diversity of results in 1899, but it was not until he reread Mendel in 1900 that he was able to select and organize his data into a rational system. Tschermak had not read Mendel before obtaining his results, and his first account of his data offers an interpretation in terms of hereditary potency. He described the 3:1 ratio as an "unequal valancy" (*Wertigkeit*). In subsequent papers he incorporated the Mendelian theory of segregation and the purity of the germ cells into his text.

In Great Britain, biologist William Bateson became the leading proponent of Mendel's theory. Around him gathered an enthusiastic band of followers. However, Darwinian evolution was assumed to be based chiefly on the selection of small, blending variations, whereas Mendel worked with clearly nonblending variations. Bateson soon found that championing Mendel aroused opposition from Darwinians. He and his supporters were called Mendelians, and their work was considered irrelevant to evolution. It took some three decades before the Mendelian theory was sufficiently developed to find its rightful place in evolutionary theory.

The distinction between a characteristic and its determinant was not consistently made by Mendel or by his successors, the early Mendelians. In 1909 Danish botanist and geneticist Wilhelm Johannsen clarified this point and named the determinants genes. Four years later American zoologist and geneticist Thomas Hunt Morgan located the genes on the chromosomes, and the popular picture of them as beads on a string emerged. This discovery had implications for Mendel's claim of an independent transmission of traits, for genes close together on the same chromosome are not transmitted independently. Moreover, as genetic studies pushed the analysis down to smaller and smaller dimensions, the Mendelian gene appeared to fragment. Molecular genetics has thus challenged any attempts to achieve a unified conception of the gene as the elementary unit of heredity. Today the gene is defined in several ways, depending upon the nature of the investigation. Genetic material can be synthesized, manipulated, and hybridized with genetic material from other species, but to fully understand its functions in the whole organism, an understanding of Mendelian inheritance is necessary. As the architect of genetic experimental and statistical analysis, Mendel remains the acknowledged father of genetics.

Premedalian Experiments

Gregor Mendel was not the first to conduct hybridization experiments, but he was the first to consider one trait at one time and this was perhaps the secret of his success. His experiments were in fact the extension and development of hybridization experiments on pea conducted by earlier workers like **Knight** (1799) and Goss (1824).

Among the workers who performed hybridization experiments before Mendel conducted his hybridization experiments on pea, the most important name is that of **J. Kolreuter** (1733-1806), a German botanist. He performed hybridization experiments in tobacco and compared the hybrids with their parents to demonstrate that the hybrids may resemble one or the other parent or may be intermediate between them. He also showed that both the parents make equal contributions to the hybrids. This he demonstrated by making **reciprocal crosses** (A@& x BB&; B@& x AB&). Another significant observation he made was, that although the hybrids themselves may be fairly uniform, their offsprings exhibit considerable diversity.

Mendel's Experiments

Mendel is known as the father of genetics because of his ground-breaking work on inheritance in pea plants 150 years ago.

Gregor Johann Mendel was a monk and teacher with interests in astronomy and plant breeding. He was born in 1822, and at 21, he joined a monastery in Brünn (now in the Czech Republic). The monastery had a botanical garden and library and was a centre for science, religion and culture. In 1856, Mendel began a series of experiments at the monastery to find out how traits are passed from generation to generation. At the time, it was thought that parents' traits were blended together in their progeny.

Studying traits in peas

Mendel studied inheritance in peas (*Pisum sativum*). He chose peas because they had been used for similar studies, are easy to grow and can be sown each year. Pea flowers contain both male and female parts, called stamen and stigma, and usually self-pollinate. Self-pollination happens before the flowers open, so progeny are produced from a single plant.

Peas can also be cross-pollinated by hand, simply by opening the flower buds to remove their pollen-producing stamen (and prevent self-pollination) and dusting pollen from one plant onto the stigma of another.

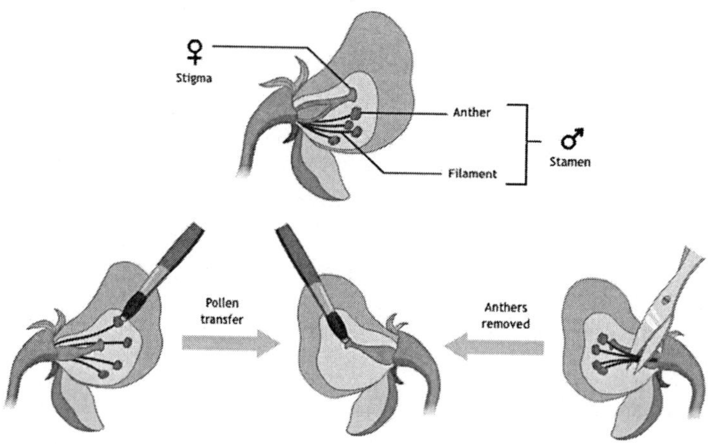

To cross-pollinate peas, pollen from the stamen of 1 plant is transferred to the stigma of another. Before the transfer, the anthers must be removed from the recipient plant to prevent self-pollination.

Traits in pea plants

Mendel followed the inheritance of 7 traits in pea plants, and each trait had 2 forms. He identified pure-breeding pea plants that consistently showed 1 form of a trait after generations of self-pollination.

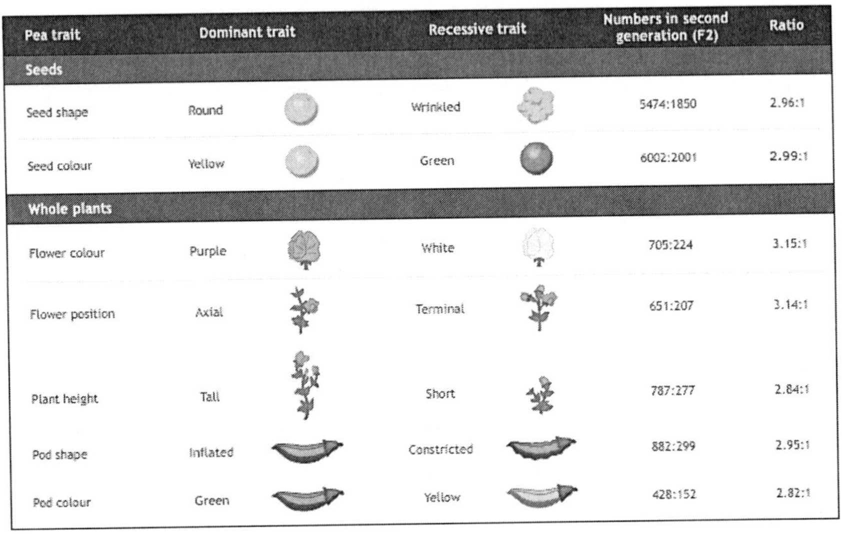

Pea trait	Dominant trait	Recessive trait	Numbers in second generation (F2)	Ratio
Seeds				
Seed shape	Round	Wrinkled	5474:1850	2.96:1
Seed colour	Yellow	Green	6002:2001	2.99:1
Whole plants				
Flower colour	Purple	White	705:224	3.15:1
Flower position	Axial	Terminal	651:207	3.14:1
Plant height	Tall	Short	787:277	2.84:1
Pod shape	Inflated	Constricted	882:299	2.95:1
Pod colour	Green	Yellow	428:152	2.82:1

Mendel cross-bred peas with 7 pairs of pure-bred traits. First-generation (F1) progeny only showed the dominant traits, but recessive traits reappeared in the self-pollinated second-generation (F2) plants in a 3:1 ratio of dominant to recessive traits.

Mendel then crossed these pure-breeding lines of plants and recorded the traits of the hybrid progeny. He found that all of the first-generation (F1) hybrids looked like 1 of the parent plants. For example, all the progeny of a purple and white flower cross were purple (not pink, as blending would have predicted). However, when he allowed the hybrid plants to self-pollinate, the hidden traits would reappear in the second-generation (F2) hybrid plants.

Dominant and recessive traits

Mendel described each of the trait variants as dominant or recessiveDominant traits, like purple flower colour, appeared in the F1 hybrids, whereas recessive traits, like white flower colour, did not.

Mendel did thousands of cross-breeding experiments. His key finding was that there were 3 times as many dominant as recessive traits in F2 pea plants (3:1 ratio).

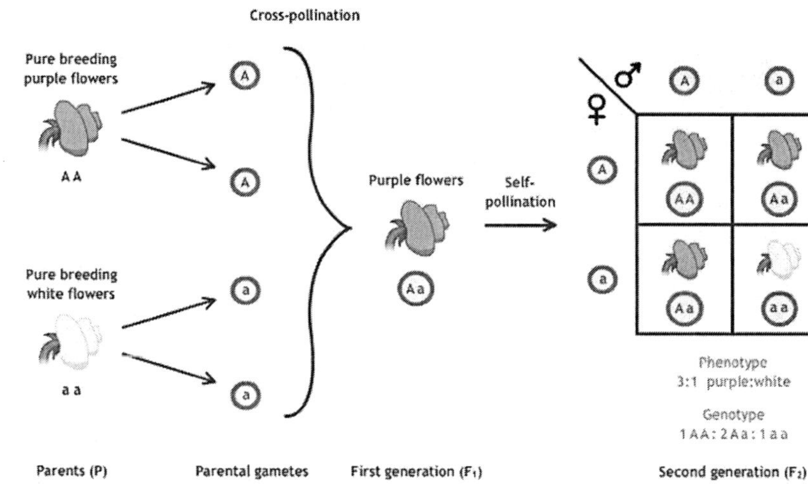

Mendel crossed pure lines of pea plants. Dominant traits, like purple flower colour, appeared in the first-generation hybrids (F1), whereas recessive traits, like white flower colour, were masked. However, recessive traits reappeared in second-generation (F2) pea plants in a ratio of 3:1 (dominant to recessive).

Traits are inherited independently

Mendel also experimented to see what would happen if plants with 2 or more pure-bred traits were cross-bred. He found that each trait was inherited independently of the other and produced its own 3:1 ratio. This is the principle of independent assortment.

Result from Mendel's experiments

Mendel's First Set of Experiments

At first, Mendel experimented with just one characteristic at a time. He began with flower colour. Mendel cross-pollinated purple- and white-flowered parent plants. The parent plants in the experiments are referred to as the P (for parent) generation.

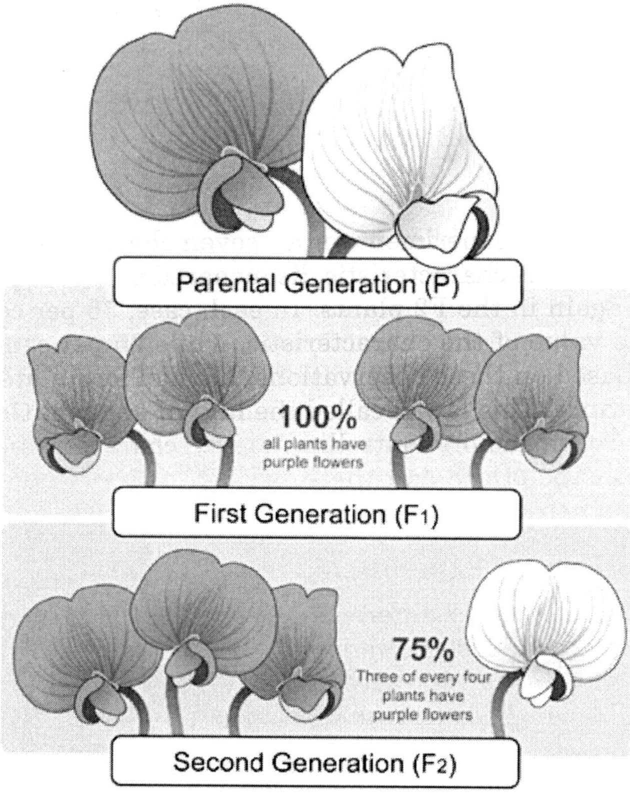

Figure : Mendel's first experiment with pea plants.

Figure: Mendel's first experiment with pea plants. The F1 generation results from the cross-pollination of two parent (P) plants, and it contains all purple flowers. The F2 generation results from the self-pollination of F1 plants, and contains 75% purple flowers and 25% white flowers.

F1 and F2 Generations

The offspring of the P generation are called the F1 (for filial, or "offspring") generation. All of the plants in the F1 generation had purple flowers — none of them had white flowers. Mendel wondered what had happened to the white-flower characteristic. He assumed that some type of inherited factor produces white flowers and some other inherited factor produces purple flowers. Did the white-flower factor just disappear in the F1 generation? If so, then the offspring of the F1 generation — called the F2 generation — should *all* have purple flowers like their parents.

To test this prediction, Mendel allowed the F1 generation plants to self-pollinate. He was surprised by the results. Some of the F2 generation plants had white flowers. He studied hundreds of F2 generation plants, and for every three purple-flowered plants, there was an average of one white-flowered plant.

Law of Segregation

Mendel did the same experiment for all seven characteristics. In each case, one value of the characteristic disappeared in the F1 plants, later showing up again in the F2 plants. In each case, 75 per cent of F2 plants had one value of the characteristic, while 25 per cent had the other value. Based on these observations, Mendel formulated his first law of inheritance. This law is called the **law of segregation**. It states that there are two factors controlling a given characteristic, one of which dominates the other, and these factors separate and go to different gametes when a parent reproduces.

Mendel's Second Set of Experiments

Mendel wondered whether different characteristics are inherited together. For example, are purple flowers and tall stems always inherited together, or do these two characteristics show up in different combinations in offspring? To answer these questions, Mendel next investigated two characteristics at a time. For example, he crossed plants with yellow round seeds and plants with green wrinkled seeds.

The outcome of a cross between plants that differ in seed colour (yellow or green) and seed form (shown here with a smooth round appearance or wrinkled appearance). The letters R, r, Y, and y represent genes for the characteristics Mendel was studying. Mendel didn't know about genes, however, because genes would not be discovered until several decades later. This experiment demonstrates that, in the F2 generation, nine out of 16 were round yellow seeds, three out of 16 were wrinkled yellow seeds, three out of 16 were round green seeds, and one out of 16 was wrinkled green seeds.

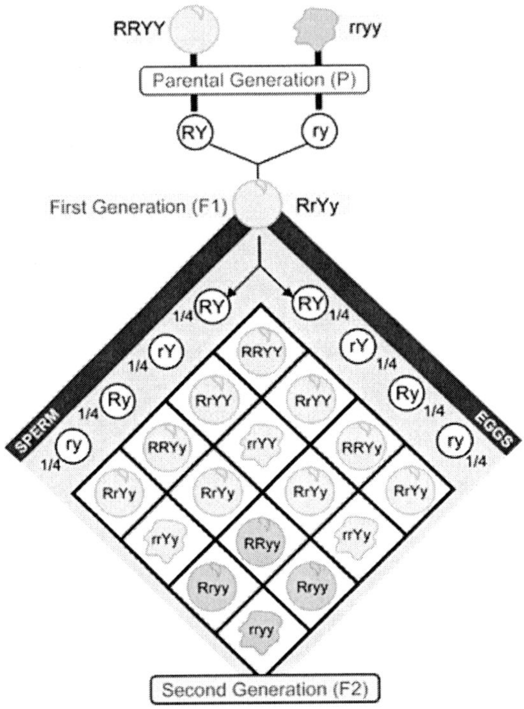

Figure : Mendel's second set of experiments.

F1 and F2 Generations

In this set of experiments, Mendel observed that plants in the F1 generation were all alike. All of them had yellow round seeds like one of the two parents. When the F1 generation plants self-pollinated, however, their offspring — the F2 generation — showed all possible combinations of the two characteristics. Some had green round seeds, for example, and some had yellow wrinkled seeds. These combinations of characteristics were not present in the F1 or P generations.

Principle of segregation or law of purity of gametes

Law of segregation is also known as the law of purity of gametes. The gametes which are formed are always pure for a particular character. A gamete may carry either the dominant or the recessive factor but not both as we find in F1 individuals. This is why it is called either as 'principle of segregation' or as 'law of purity of gametes'.

The law of segregation is also called the law of purity of gametes. According to this law, the two alleles of a gene separate during formation of gametes. They do not blend but segregate or separate into different gametes. Hence each gamete is "pure" and possesses only one allele of a gene. The gametes combine together by random fusion at the time of fertilisation.

The law of dominance is the first law. It states that characters are controlled by discrete units called factors. Factors occur in pairs. In a dissimilar pair of factors one member of the pair dominates (dominant) the other (recessive).

Law of independent assortment states that when two or more characteristics are inherited as in a dihybrid cross, individual hereditary factors assort independently during gamete production, giving different traits an equal opportunity of occurring together.

Principle of Segregation and its Importance

The principle of segregation defined that the individual has two alleles for each particular characteristic, and during the development of gametes, these alleles become segregated. In other words, there is one allele in each gamete. The principle of segregation is vital because it describes how genotypic ratios are produced in the haploid gametes.

Where does the law of segregation occur in meiosis?

Mendel's Segregation law occurs in anaphase (I and II) of meiosis. It is a phase in the first meiotic division in which the homologous chromosomes are segregated into two daughter nuclei with their various versions of each gene. During meiosis, the behavior of homologous chromosomes can contribute to the separation of the alleles into distinct gametes for each genetic locus. When chromosomes divide during meiosis into various gametes, the two distinct alleles for a single gene often segregate such that one of the two alleles is obtained by each gamete.

Why is the Law of Segregation universally accepted?

The Law of Segregation is a universally accepted law of inheritance because it is the only inheritance law that has no exceptions while the other two laws have some exceptions. It states that each gene consisting of two alleles that differ during the development of gametes, one allele from both mother and father, combines during fertilization.

Law of Independent Assortment

Also called the second inheritance law, it states that:

"Separate couples with alleles are transferred separately from each

other to the next generation. As a result, gene inheritance does not influence gene inheritance somewhere else at one position in the genome."

OR

"This law described that alleles of various genes that are distributed during gamete development assort independently of each other."

This law is valid for those traits that are not related to each other such as seed color and seed shape. When an individual inherits two or more characteristics, those characteristics are assorted independently during the production of gamete. This gives the different traits an equivalent probability of occurring together. This indicates that the inheritance of one character will not influence the inheritance of the other.

Explanation

When two sets of Mendelian traits are fused into a hybrid, one pair of traits differs from the other pair of traits. Therefore, it means the alleles are independent and do not affect the other alleles. For example, a pea plant with round and yellow shape seeds was cross-pollinated with a plant with wrinkled green shape seeds.

When does this process occur?

While crossing over occurs in Prophase I, independent assortment law can be observed during metaphase I and anaphase I of meiosis. In metaphase, for instance, the chromosomes line up along the metaphase plate in random orientation.

During meiosis, gamete cells are the final product. Gamete cells are referred to as haploid cells and also have half the regular diploid cell DNA. It is a vital aspect of reproduction that enables the cells of gametes to fuse to form a diploid zygote, carrying the information of DNA that is required for the development of offspring and a chromosomal number that is maintained across generations.

Principle of Independent Assortment

Independent assortment principles describe that during the development of gametes, allele pairs are segregated, which means that the traits are passed to the offspring independently of one another.

Importance of Law of independent assortment

It is important for different genetic variations in organisms. For instance, the gene or alleles coding for a trait segregates independently from the gene or alleles coding for another trait during the development of gametes. It is also essential for the production of new genetic variations that enhance the genetic diversity within the population.

Law of Purity of Gametes

Definition

Gregor Mendel's law stating that the *unit factors* are brought together and paired during union or fertilization; however, these unit factors do not affect or mix with each other though they have become paired.

Supplement

Breeding and testing about 5,000 pea plants, he was able to come up with crucial generalizations that were later on used as founding principles of the Mendelian inheritance or Mendel's Principles of Heredity. The so-called "Mendel's laws of inheritance" came about based on the set of principles of Gregor Mendel. These laws are the *Law of Segregation*, the *Law of Independent Assortment*, the *Law of Dominance*, and the *Law of Unit Characters*.

Unit factors from maternal and paternal gametes combine together and pair during the union of gametes (fertilization). However, the bringing together and pairing of these unit factors do not eventuate to their blending, altering, or mixing. They remain *pure* in a way that they remain as distinct entities within the somatic cells of the offspring. When the offspring does form its own gametes, the members of the paired unit factors would separate or segregate. Hence, this law is also referred to as Law of Segregation. Mendel further assumes that each of them assort independently into the newly formed gametes.

Principle of independent Assortment (second law)

The Principle of Independent Assortment describes how different genes independently separate from one another when reproductive cells develop. Independent assortment of genes and their corresponding traits was first observed by Gregor Mendel in 1865 during his studies of genetics in pea plants. Mendel was performing dihybrid crosses, which are crosses between organisms that differ with regard to two traits. He discovered that the combinations of traits in the offspring of his crosses did not always match the combinations of traits in the parental organisms. From his data, he formulated the Principle of Independent Assortment.

We now know that this independent assortment of genes occurs during meiosis in eukaryotes. Meiosis is a type of cell division that reduces the number of chromosomes in a parent cell by half to produce four reproductive cells called gametes. In humans, diploid cells contain 46 chromosomes, with 23 chromosomes inherited from the mother and a second similar set of 23 chromosomes inherited from the father. Pairs of similar chromosomes are called homologous chromosomes. During meiosis, the pairs of homologous chromosome are divided in half to form haploid cells, and this separation, or assortment, of homologous chromosomes is random. This means that all of the maternal chromosomes will not be separated into one cell, while the all paternal chromosomes are separated into another. Instead, after meiosis occurs, each haploid cell contains a mixture of genes from the organism's mother and father.

Another feature of of independent assortment is recombination. Recombination occurs during meiosis and is a process that breaks and recombines pieces of DNA to produce new combinations of genes. Recombination scrambles pieces of maternal and paternal genes, which ensures that genes assort independently from one another. It is important to note that there is an exception to the law of independent assortment for genes that are located very close to one another on the same chromosome because of genetic linkage.

Dihybrid Cross, Trihybrid cross, trihybrid test cross

Dihybrid Cross and Inheritance of Two Genes

"Dihybrid cross is the cross between two different genes that differ in two observed traits."

Gregor Johann Mendel was the first person who discovered the basic principles of heredity during the mid-19th century. Hence, he is known as the "Father of Modern Genetics". He conducted experiments in his garden on pea plants and observed their pattern of inheritance from one generation to the next generation.

Mendel laid the basic groundwork in the field of genetics and eventually proposed the laws of inheritance. Law of Segregation, Law of Independent Assortment and Law of Dominance are the three laws of inheritance proposed by Gregor Mendel. These laws came into existence from his experiments on pea plants with a variety of traits.

Mendel first studied the inheritance of one gene in the plant through monohybrid cross. He considered only a single character (plant height) on pairs of pea plants with one contrasting trait. Later, he studied the inheritance of two genes in the plant through dihybrid cross.

Mendel studied the following seven characters with contrasting traits:

- Stem height: Tall/dwarf
- Seed shape: Round/wrinkled
- Seed colour: Yellow/green
- Pod colour: Green/yellow
- Pod shape: Inflated/constricted
- Flower colour: Violet/white
- Flower position: Axial/terminal

Dihybrid Cross

A dihybrid cross is a breeding experiment between two organisms which are identical hybrids for two traits. In other words, a dihybrid cross is a cross between two organisms, with both being heterozygous for two different traits. The individuals in this type of trait are homozygous for a specific trait. These traits are determined by DNA segments called genes.

In a dihybrid cross, the parents carry different pair of alleles for each trait. One parent carries homozygous dominant allele, while the other one carries homozygous recessive allele. The offsprings produced after the crosses in the F1 generation are all heterozygous for specific traits.

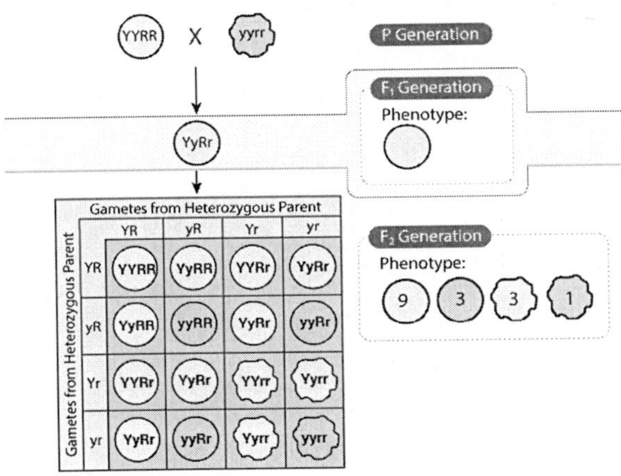

Visualisation of dihybrid cross using a Punnett square

Dihybrid Cross Examples

Mendel took a pair of contradicting traits together for crossing, for example colour and the shape of seeds at a time. He picked the wrinkled-green seed and round-yellow seed and crossed them. He obtained only round-yellow seeds in the F1 generation. This indicated that round shape and yellow colour of seeds are dominant in nature.

Meanwhile, the wrinkled shape and green colour of seeds are recessive traits. Then, F1 progeny was self-pollinated. This resulted in four different combinations of seeds in the F2 generation. They were wrinkled-yellow, round-yellow, wrinkled-green seeds and round-green in the phenotypic ratio of 9:3:3:1.

During monohybrid cross of these traits, he observed the same pattern of dominance and inheritance. The phenotypic ratio 3:1 of yellow and green colour and of round and wrinkled seed shape during monohybrid cross was retained in dihybrid cross as well.

Consider "Y" for yellow seed colour and "y" for green seed colour, "R" for round shaped seeds and "r" for wrinkled seed shape. Thus, the parental genotype will be "YYRR" (yellow-round seeds) and "yyrr" (green-wrinkled seeds).

Trihybrid cross:

1. With three unlinked genes, each parent can produce 8 different types of gametes, which generates 64 possible genotypic combinations.
2. The phenotypic ratio of the trihybrid cross is 27:9:9:9:3:3:3:1.
3. These are the phenotypes or the visible characters of the offsprings at the end of trihybrid cross.

Analysis of a tri-hybrid test cross

Theoretical:

Given three loci, linked on a single chromosome in the following order

——-*A*————*B*————*C*——

Then, a cross between **ABC//ABC** x **abc//abc tester** involves

——A————B————C——

——**A**————**B**————**C**——-

——a————b————c——

——**a**————**b**————**c**——-

Recombination may *not* occur at all (**parental type**),

or may occur between **A & B**, or **B & C** (**single crossovers**)

or between **both**: a **double crossover**

During **Meiosis I**, single cross-over between the **A & B** loci looks like:

```
——A——————B—————C——- => ABC Parental
——A—\ /—B—————C——- => aBC Single recombinant
        X
——a—/ \—b—————c——- => Abc Single recombinant
——a——————b—————c——- => abc Parental
```

A single cross-over between the **B & C** loci looks like:

```
——A—————B——————C——- => ABC Parental
——A—————B—\ /—C——- => abC Single recombinant
            X
——a—————b—/ \—c——- => ABc Single recombinant
——a—————b——————c——- => abc Parental
```

Each crossover ***changes the phase*** of one locus with respect to (wrt) its nearest neighbor.

The frequency of either single crossover is proportional to the distance between loci, and *increases* with distance

The frequency of a **double crossover** is the *product* of these frequencies:

therefore a double-crossover is the *rarest* type:

```
——A——————B——————C——
——A—\ /—B—\ /—C——
       X      X
——a—/ \—b—/ \—c——
——a——————b——————c——-
```

A double crossover will ***reverse the phase*** of the *middle* locus

```
——A——————B——————C——- => ABC
——A——————b——————C——- => AbC
——a——————B——————c——- => aBc
——a——————b——————c——- => abc
```

Thus, when you compare the *most common* types (the **parental ABC** & **abc**)

with the *rarest* (the **double-recombinants AbC** & **aBc**)

the ***cis / trans*** **phase** of the ***middle*** locus is *reversed* wrt both ***outside*** loci

Here: **B** is reversed wrt **A & C** (and **b** wrt **a & c**): ***B*** *is the middle locus*

Experimental

Consider three loci **E, F** & **G**, linked on a single chromosome, *but in an unknown order*

The following cross is constructed:

EEFFGG x **eeffgg** **(P)** => **EeFfGg** x **eeffgg** (tester) ($\mathbf{F_1}$)

which produces the following counts among 1,000 $\mathbf{F_2}$ offspring.

Use of a tester means the genotype of the offspring can be inferred directly from the phenotype

Genotype	Phenotype	Count	Class
E F G efg	**'EFG'**	370	**P**
E F g efg	**'EFg'**	8	**D**
E f G efg	**'EfG'**	37	**II**
E f g efg	**'Efg'**	95	**I**
e F G efg	**'eFG'**	85	**I**
e F g efg	**'eFg'**	43	**II**
e f G efg	**'efG'**	12	**D**
e f g efg	**'efg'**	350	**P**

(1) Identify the **four pairs** of phenotype classes:

one **parental (P)** class, which is most frequent, and which shows the same phase relationships as the parents

two **single recombinant (I & II)** classes, between the middle locus and each outside locus

one **double recombinant (D)** class, between the middle locus and both outside loci

As this requires two events, this is the rarest class

The two members of each recombinant pair should occur in approximately equal numbers, because each crossover produces two products

(2) To determine **gene locus order**:

Most common phenotypes **EFG** & **efg** are **parental: no recombinants**

Rarest phenotype **EFg** & **efG** will be the **double-recombinants**:

Compare phases of [**EFG efg**] vs [**EFg efG**]

* The ***G*** locus *changes phase* ("*flips*") wrt **E & F**, therefore ***G* is in the middle** *

——*-E——G——F*——-

(3) To determine the **frequency of recombination** between locus pairs,

count all classes in which the phase changes from the parental types

(i) for **E & G**

EG & eg are parental types

Eg & eG are **recombinant** types between ***E & G***: (Class I):

count all classes with this type

\# **(EFg + efG + eFG + Efg)**

[(8 + 12) + (95 + 85)] / 1000 = **0.20 = 20% = 20 m.u. = 20 cM**

(ii) for **G & F**

GF & gf are parental types

Gf & gF are recombinant types between ***F & G*** (Class II):

(**EFg** + **efG** + **EfG** + **eFg**)

[(8 + 12) + (37 + 43)] / 1000 = **0.10 = 10% = 10 m.u. = 10 cM**

(iii) for **E & F**

EF & ef are parental types

Ef & eF are recombinant types between ***E & F***

(**EfG** + **eFg** + **eFG** + **Efg**)

[(37 + 43) + (95 + 85)] / 1000 = **0.26 = 26% = 26m.u. = 26 cM**

(4) **Map data** then look like this:

The distance between **E & F** is greatest, so these are on the outside

E————20cM————G—-10cM—-F

E—————-26cM—————-F

but **0.26 < (0.20 + 0.10)** ***WHY***?

In the double recombinants,

—A——B——C— ⇒ **ABC**

—A——b——C— ⇒ **AbC**

—a——B——c— ⇒ **aBc**

—a——b——c— ⇒ **abc**

so ***double recombinants resemble parental type***

but each represents **2 cross-overs** between outside (***E & F***) loci

(3.iii') Then the ***corrected count*** of crossovers between **E & F**

[(37 + 43) + (95 + 85) + **(2)(8 + 12)**] = 0.3 = 30 m.u. = **30 cM**

(5) **Interference** occurs where one cross-over *decreases* the probability of others nearby:

Compare the **observed** *vs* **expected** # double-recombinants

Interference (I) = 1 - (observed # D) / (expected # D) = 1 - [(8 + 12) / (**0.20**)(**0.10**)(1000)] = 1 - $\mathbf{f(D_{obs})}$ / $\mathbf{f(D_{exp})}$ = 1 - [(0.008 + 0.012) / [(0.20))(0.10)]

In these data, **I** = 1 - (8+12)/(20) = **0** : # expected = # observed ⇒ no interference

Mendal's Result - chromosome theory and linkage

Long before chromosomes were visualized under a microscope, the father of modern genetics, Gregor Mendel, began studying heredity in 1843. With the improvement of microscopic techniques during the late 1800s, cell biologists could stain and visualize subcellular structures with dyes and observe their actions during cell division and meiosis. With each mitotic division, chromosomes replicated, condensed from an amorphous (no constant shape) nuclear mass into distinct X-shaped bodies (pairs of identical sister chromatids), and migrated to separate cellular poles.

Chromosomal Theory of Inheritance

The speculation that chromosomes might be the key to understanding heredity led several scientists to examine Mendel's publications and re-evaluate his model in terms of the behavior of chromosomes during mitosis and meiosis. In 1902, Theodor Boveri observed that proper embryonic development of sea urchins does not occur unless chromosomes are present. That same year, Walter Sutton observed the separation of chromosomes into daughter cells during meiosis ([link]). Together, these observations led to the development of the **Chromosomal Theory of Inheritance**, which identified chromosomes as the genetic material responsible for Mendelian inheritance.

(a) Walter Sutton and (b) Theodor Boveri are credited with developing the Chromosomal Theory of Inheritance, which states that chromosomes carry the unit of heredity (genes).

Genetic Linkage and Distances

Mendel's work suggested that traits are inherited independently of each other. Morgan identified a 1:1 correspondence between a segregating trait and the X chromosome, suggesting that the random segregation of chromosomes was the physical basis of Mendel's model. This also demonstrated that linked genes disrupt Mendel's predicted outcomes. The fact that each chromosome can carry many linked genes explains how individuals can have many more traits than they have chromosomes. However, observations by researchers in Morgan's laboratory suggested that alleles positioned on the same chromosome were not always inherited together. During meiosis, linked genes somehow became unlinked.

Homologous Recombination

In 1909, Frans Janssen observed chiasmata—the point at which chromatids are in contact with each other and may exchange segments—prior to the first division of meiosis. He suggested that alleles become unlinked and chromosomes physically exchange segments. As chromosomes condensed and paired with their homologs, they appeared to interact at distinct points. Janssen suggested that these points corresponded to regions in which chromosome segments were exchanged. It is now known that the pairing and interaction between homologous chromosomes, known as synapsis, does more than simply organize the homologs for migration to separate daughter cells. When synapsed, homologous chromosomes undergo reciprocal physical exchanges at their arms in a process called **homologous recombination**, or more simply, "crossing over."

To better understand the type of experimental results that researchers were obtaining at this time, consider a heterozygous individual that inherited dominant maternal alleles for two genes on the same chromosome (such as *AB*) and two recessive paternal alleles for those same genes (such as *ab*). If the genes are linked, one would expect this individual to produce gametes that are either *AB* or *ab* with a 1:1 ratio. If the genes are unlinked, the individual should produce *AB*, *Ab*, *aB*, and *ab* gametes with equal frequencies, according to the Mendelian concept of independent assortment. Because they correspond to new allele combinations, the genotypes Ab and aB are **nonparental types** that result from homologous recombination during meiosis. **Parental types** are progeny that exhibit the same allelic combination as their parents. Morgan and his colleagues, however, found that when such heterozygous individuals were test crossed to a homozygous recessive parent (*AaBb* × *aabb*), both parental and nonparental cases occurred. For example, 950 offspring might be recovered that were either *AaBb* or *aabb*, but 50 offspring would also be obtained that were either *Aabb* or *aaBb*. These results suggested that linkage occurred most often, but a significant minority of offspring were the products of recombination.

Art Connection

Inheritance patterns of unlinked and linked genes are shown. In (a), two genes are located on different chromosomes so independent assortment occurs during meiosis. The offspring have an equal chance of being the parental type (inheriting the same combination of traits as the parents) or a nonparental type (inheriting a different combination of traits than the parents). In (b), two genes are very close together on the same chromosome so that no crossing over occurs between them. The genes are therefore always inherited together and all of the offspring are the parental type. In (c), two genes are far apart on the chromosome such that crossing over occurs during every meiotic event. The recombination frequency will be the same as if the genes were on separate chromosomes. (d) The actual recombination frequency of fruit fly wing length and body color that Thomas Morgan observed in 1912 was 17 percent. A crossover frequency between 0 percent and 50 percent indicates that the genes are on the same chromosome and crossover occurs some of the time.

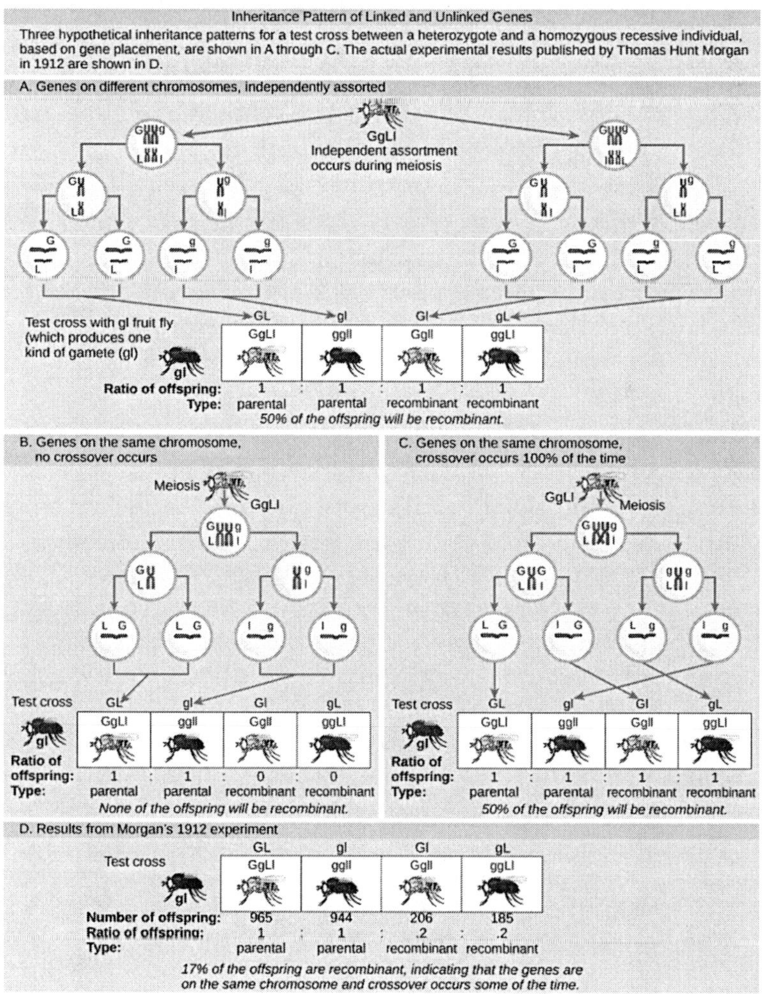

In a test cross for two characteristics such as the one shown here, can the predicted frequency of recombinant offspring be 60 percent? Why or why not?

Genetic Maps

Janssen did not have the technology to demonstrate crossing over so it remained an abstract idea that was not widely accepted. Scientists thought chiasmata were a variation on synapsis and could not understand how chromosomes could break and rejoin. Yet, the data were clear that linkage did not always occur. Ultimately, it took a young undergraduate student and an "all-nighter" to mathematically elucidate the problem of linkage and recombination.

In 1913, Alfred Sturtevant, a student in Morgan's laboratory, gathered results from researchers in the laboratory, and took them home one night to mull them over. By the next morning, he had created the first "chromosome map," a linear representation of gene order and relative distance on a chromosome.

By using recombination frequency to predict genetic distance, the relative order of genes on chromosome 2 could be inferred. The values shown represent map distances in centimorgans (cM), which correspond to recombination frequencies (in percent). Therefore, the genes for body color and wing size were 65.5 " 48.5 = 17 cM apart, indicating that the maternal and paternal alleles for these genes recombine in 17 percent of offspring, on average.

To construct a chromosome map, Sturtevant assumed that genes were ordered serially on threadlike chromosomes. He also assumed that the incidence of recombination between two homologous chromosomes could occur with equal likelihood anywhere along the length of the chromosome. Operating under these assumptions, Sturtevant postulated that alleles that were far apart on a chromosome were more likely to dissociate during meiosis simply because there was a larger region over which recombination could occur. Conversely, alleles that were close to each other on the chromosome were likely to be inherited together. The average number of crossovers between two alleles—that is, their **recombination frequency**—correlated with their genetic distance from each other, relative to the locations of other genes on that chromosome. Considering the example cross between *AaBb* and *aabb* above, the frequency of recombination could be calculated as 50/1000 = 0.05. That is, the likelihood of a crossover between genes *A/a* and *B/b* was 0.05, or 5 percent. Such a result would indicate that the genes were definitively linked, but that they were far enough apart for crossovers to occasionally occur. Sturtevant divided his genetic map into map units, or **centimorgans (cM)**, in which a recombination frequency of 0.01 corresponds to 1 cM.

By representing alleles in a linear map, Sturtevant suggested that genes can range from being perfectly linked (recombination frequency = 0) to being perfectly unlinked (recombination frequency = 0.5) when genes are on different chromosomes or genes are separated very far apart on the same chromosome. Perfectly unlinked genes correspond to the frequencies predicted by Mendel to assort independently in a dihybrid cross. A recombination frequency of 0.5 indicates that 50 percent of offspring are recombinants and the other 50 percent are parental types. That is, every type of allele combination is represented with equal frequency. This representation allowed Sturtevant to additively calculate distances between several genes on the same chromosome. However, as the genetic distances approached 0.50, his predictions became less accurate because it was not clear whether the genes were very far apart on the same chromosome or on different chromosomes.

In 1931, Barbara McClintock and Harriet Creighton demonstrated the crossover of homologous chromosomes in corn plants. Weeks later, homologous recombination in *Drosophila* was demonstrated microscopically by Curt Stern. Stern observed several X-linked phenotypes that were associated with a structurally unusual and dissimilar X chromosome pair in which one X was missing a small terminal segment, and the other X was fused to a piece of the Y chromosome. By crossing flies, observing their offspring, and then visualizing the offspring's chromosomes, Stern demonstrated that every time the offspring allele combination deviated from either of the parental combinations, there was a corresponding exchange of an X chromosome segment. Using mutant flies with structurally distinct X chromosomes was the key to observing the products of recombination because DNA sequencing and other molecular tools were not yet available. It is now known that homologous chromosomes regularly exchange segments in meiosis by reciprocally breaking and rejoining their DNA at precise locations.

Mendel's Mapped Traits

Homologous recombination is a common genetic process, yet Mendel never observed it. Had he investigated both linked and unlinked genes, it would have been much more difficult for him to create a unified model of his data on the basis of probabilistic calculations. Researchers who have since mapped the seven traits investigated by Mendel onto the seven chromosomes of the pea plant genome have confirmed that all of the genes he examined are either on separate chromosomes or are sufficiently far apart as to be statistically unlinked. Some have suggested that Mendel was enormously lucky to select only unlinked genes, whereas others question whether Mendel discarded any data suggesting linkage. In any case, Mendel consistently observed independent assortment because he examined genes that were effectively unlinked.

Molecular basis of characters studies Mendel

MODERN genetics is founded on the classical study of seven genes by Gregor Mendel, even though until relatively recently little was known about their function. Recent advances in molecular biology have dramatically changed this situation and unveiled the identity of four of Mendel's genes. Mendel started his experiments by carefully selecting seven qualitative characters in peas (*Pisum sativum L.*) in 1856, and his results were published in 1866. His work then went largely unnoticed for the next 34 years before its rediscovery by three independent researchers—Hugo de Vries, Carl Correns, and Erich von Tschermak—in 1900 (Bateson 1900). However, as soon as the work was rediscovered, it created controversy (*e.g.*, Weldon 1902). The closeness of Mendel's experimental observations to those predicted by his theories have led to numerous articles and ongoing debate about whether the data could have been obtained in the published form without some modification. There have been many plausible arguments made for and against this view by a range of eminent statisticians and geneticists (*e.g.*, Fisher 1936; Fairbanks and Rytting 2001; Hartl and Fairbanks 2007). Some have gone so far as to suggest that the theories behind Mendel's two laws were not even correctly articulated in his original paper (Monaghan and Corcos 1993). However, given the lack of suitable terminology at the time, this seems a very harsh judgment and one with which we do not concur. Fisher (1936) certainly saw the significance of Mendel's contribution to the field even though he was the one who raised several concerns about the data. Some of these concerns still exist, and the controversy has not been resolved, despite recent implications to the contrary (Franklin 2008; Stigler 2008). We do not intend to pursue the matter here, but we believe that rational explanations have been given to potentially explain some of the problems with the data and their interpretation (*e.g.*, Hartl and Fairbanks 2007; Franklin 2008). However, these issues cannot be definitively resolved, and it is probably time to concentrate on the remarkable understandings that arise from Mendel's insightful paper.

It is fascinating to trace the fate of Mendel's genes over the past 145 years. In some cases, we can do this with considerable certainty, tracing it through to the molecular cloning of the gene concerned and the establishment of its function, but for other genes the history is somewhat murkier. For at least one of the characters, we cannot even be sure which locus Mendel studied, as modern molecular biology has, over the last few decades, highlighted the existence of multi-gene families for some of his characters. In this sense, it is ironic that a discipline that owes so much to Mendel's work has somewhat muddied the waters regarding the genes/characters that he studied. Unfortunately, the lines of peas that Mendel used are not known with certainty (Bhattacharyya *et al.* 1990; Hellens *et al.* 2010). However, given that his research built on the previous work of others and that we have some historical accounts of what mutants were available at the outset of his experiments in the 1850s (*e.g.*, Knight 1799), we can make an educated guess in most instances.

Even harder than defining which seven loci Mendel studied is the question of which mutations he used. Theoretically, it is now quite clear that there are many possible mutations in each gene that could produce the same phenotype, especially if Mendel's original mutation was a null. Again, the only way of determining this is to see what material would have been available in central Europe at the time that Mendel performed the studies. Many of the mutations that he examined were being used in the cultivars grown at that time. Indeed, several of them had been exploited for agricultural purposes for many hundreds of years (*e.g.*, white flowers, dwarf habit, sugar pods), and some cultivars possessing these mutations that predate his work are still available in collections (*e.g.*, the John Innes *Pisum* germplasm collection).

Molecular basis of Mendel's wrinkled seed character

In the year 1990, a group of scientists, working at John Innes Institute at Norwich (U.K.), cloned the pea gene *r (rugosus),* which determines whether the seed is round or wrinkled. It was shown that an isoform of **'starch branching enzyme'** (SBE I)is present in round *(RR* or *Rr)* seeds, but absent in wrinkled *(rr)* seeds. By carefully planned experiments it was shown that there was 100% co-segregation of the absence of SBE I with *r* locus, suggesting that the gene for SBE I is located at *r* locus. It was also shown that in *rr* lines, SBE I gene is interrupted by a small DNA sequence (0.8 kilobases) called **'transposon-like insertion'**, similar to *Ac/Ds* system in maize. Interruption of SBE I gene by insertion of a foreign sequence, gives an aberrant SBE I enzyme leading to metabolic disturbances in the biosynthesis of starch, lipid and protein. This results in an increase of free sugars due to failure of starch formation. This probably leads to higher osmotic pressure and hence higher water content and larger cell volume earlier observed in *rr* seeds. The seeds lose large proportion of their volume on maturation leading to shrinkage in volume. Since testa does not shrink with the cotyledons, seeds become wrinkled.

In most cases of the study of Mendelian traits, we seldom know the molecular basis, so that the students do not ask the question : how does a gene control the phenotype? With the availability of recent techniques of molecular biology and recombinant DNA, many traits are now being examined for the molecular basis of their development. This approach is receiving support from the recent emergence of **'reverse genetics'** (from DNA to phenotype) as opposed, to **forward** or **classical genetics** (from phenotype to DNA).

Mendelian genetics in humans finding form Mendelism

Mendelian Inheritance in Humans

Characteristics that are encoded in DNA are called **genetic traits.** Different types of human traits are inherited in different ways. Some human traits have simple inheritance patterns like the traits that Gregor Mendel studied in pea plants. Other human traits have more complex inheritance patterns.

Mendelian inheritance refers to the inheritance of traits controlled by a single gene with two alleles, one of which may be dominant to the other. Not many human traits are controlled by a single gene with two alleles, but they are a good starting point for understanding human heredity. How Mendelian traits are inherited depends on whether the traits are controlled by genes on autosomes or the X chromosome.

Autosomal Traits

Autosomal traits are controlled by genes on one of the 22 human autosomes. Consider earlobe attachment. A single autosomal gene with two alleles determines whether you have attached earlobes or free-hanging earlobes. The allele for free-hanging earlobes (*F*) is dominant to the allele for attached earlobes (*f*). Other single-gene autosomal traits include widow's peak and hitchhiker's thumb. The dominant and recessive forms of these traits. Which form of these traits do you have? What are your possible genotypes for the traits?

It shows how the earlobe trait was passed from generation to generation within a family. Pedigrees are useful tools for studying inheritance patterns.

Pedigree for Earlobe Attachment

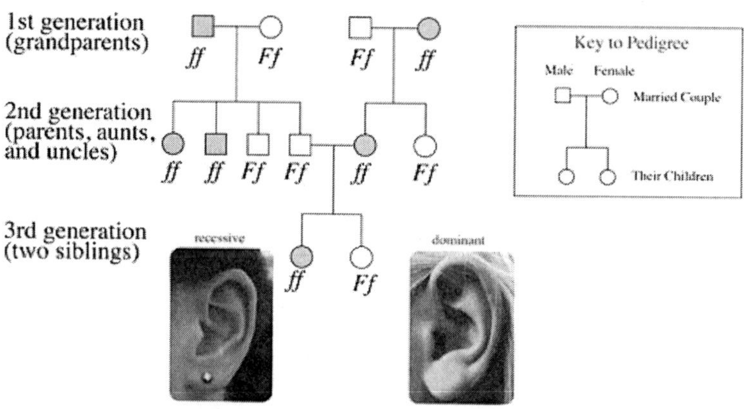

Having free-hanging earlobes is an autosomal dominant trait. This figure shows the trait and how it was inherited in a family over three generations. Shading indicates people who have the recessive form of the trait. Look at (or feel) your own earlobes. Which form of the trait do you have? Can you tell which genotype you have?

Other single-gene autosomal traits include widow's peak and hitchhiker's thumb. The dominant and recessive forms of these traits. Which form of these traits do you have? What are your possible genotypes for the traits?

Single Gene Autosomal Traits

Widow's peak

No widow's peak

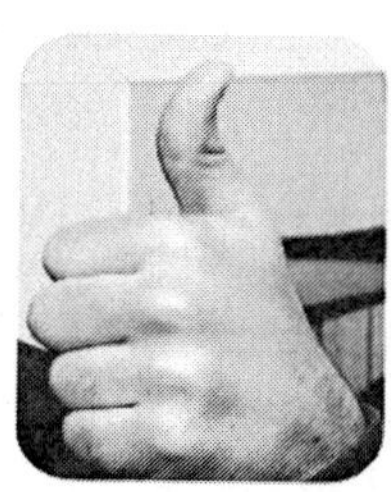
Hitchhiker's thumb

No hitchhiker's thumb

Widow's peak and hitchhiker's thumb are dominant traits controlled by a single autosomal gene.

Mendel's Experiments

Gregor Mendel performed breeding experiments in his garden to analyse patterns of inheritance. He opted for cross-bred normal pea plants with selective traits over various generations. When two plants were crossed that differed in a single trait (round peas vs. wrinkled seeds, short stems vs. tall stems, white flowers vs. purple flowers, etc), Mendel found that the next generation, F1 comprised of whole individuals that exhibit only one trait. However, after the generation was interbred, its offspring which is the F2 generation showed a 3:1 ratio wherein three individuals had similar traits like a parent.

Mendel theorized that genes could be formed by three possible combinations of heredity units that are said to be factors: AA, aa, Aa. The big 'A' shows the dominant factor and the small 'a' shows the recessive factor. The beginning plants were homozygous AA or aa, F1 generation was Aa and F2 generation was AA, aa or Aa. The interaction between these two finds the physical trait that is visible.

According to Mendel's law of Dominance, when two organisms of separate traits are crossed, every offspring shows the trait of only one dominant character. The recessive trait is expressed phenotypically only if both factors are recessive.

Studying Inheritance Patterns

There are two very useful tools for studying how traits are passed from one generation to the next. One tool is a pedigree, and the other is a Punnett square.

Pedigree

A pedigree shows how a trait is passed from generation to generation within a family. A pedigree can show, for example, whether a Mendelian trait is an autosomal or X-linked trait. It can also be used to infer the genotype of different members of the family.

The trait represented by this chart is a hypothetical autosomal trait controlled by a dominant allele. At the top of the pedigree, you can see symbols representing a married couple. The husband *has* the trait (affected male), but the wife does not (unaffected female). The next row of the pedigree shows the couple's children, as well as the spouses of three of the children. For example, the first child on the left is an affected male married to an unaffected female. The third row of the pedigree shows the next generation (the grandchildren of the couple at the top of the pedigree). One child in this generation — the affected female on the left is the sibling of an unaffected male.

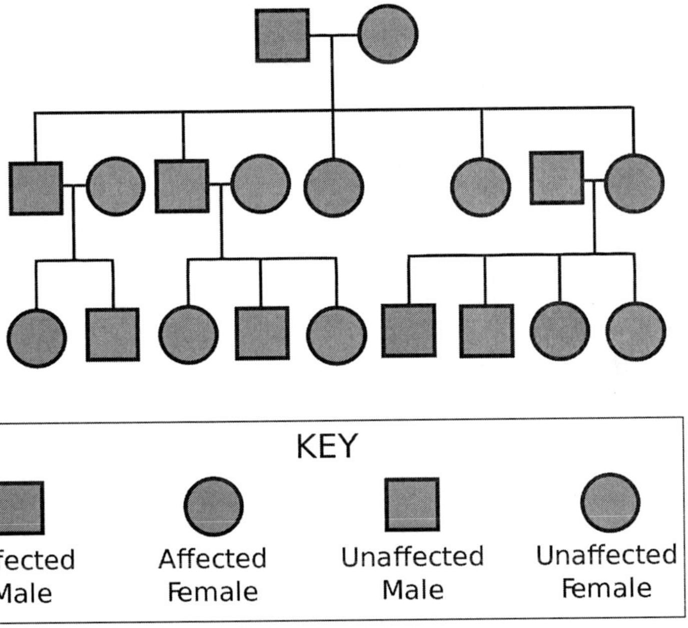

Figure : A pedigree chart is similar to a family tree. It shows how a trait is passed from parents to offspring in a family. The trait represented by this pedigree is an autosomal dominant trait.

Punnett Square

A **Punnett square** is a chart that allows you to easily determine the expected ratios of possible genotypes in the offspring of two parents. You can see a hypothetical. In this case, the gene is autosomal, and both parents are heterozygotes *(Aa)* for the gene. Half of the gametes produced by each parent will have the *A* allele, and half will have the *a* allele. That's because the two alleles are on homologous chromosomes, which always separate and go to separate gametes during meiosis. The alleles in the gametes from each parent are written down the side and across the top of the Punnett square. Filling in the cells of the Punnett square gives the possible genotypes of their children. It also shows the most likely ratios of the genotypes, which in this case is 25 per cent *AA,* 50 per cent *Aa,* and 25 per cent *aa.*

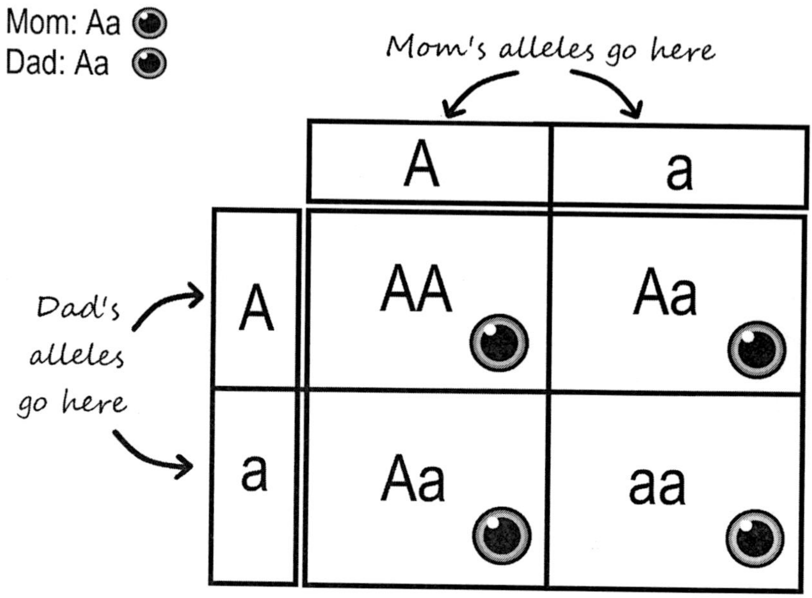

Figure : A Punnett square shows the most likely proportions of offspring by genotype for a particular mating type.

A Punnett square can also be used to show how the X and Y chromosomes are passed from parents to their children. This is illustrated in the Punnett square. It may help you understand the inheritance pattern of sex-linked traits.

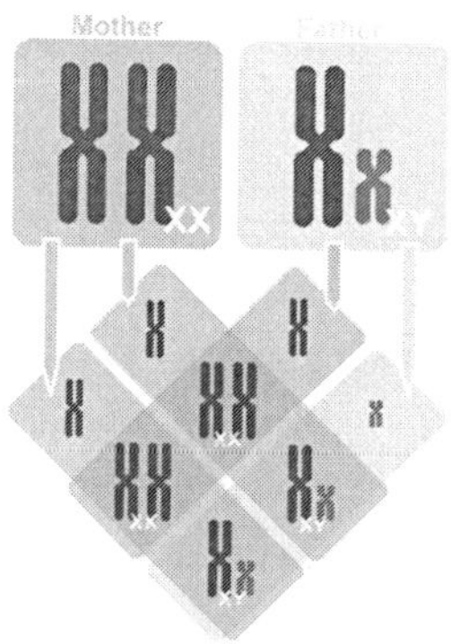

Figure : Inheritance of Sex Chromosomes. Mothers pass only X chromosomes to their children. Fathers always pass their X chromosome to their daughters, and their Y chromosome to their sons. Can you explain why fathers always determine the sex of the offspring?

Here's how to fill out a Punnett Square:

Learn Biology: How to Draw a Punnett Square, mahalodotcom, 2011.

Try out this Punnett Square:

Autosomal Mendelian Traits in Humans

Not many human autosomal traits are controlled by a single gene with two alleles, but they are a good starting point for understanding human heredity. As discussed in the beginning of this concept, most forms of albinism in humans have a Mendelian inheritance pattern. Albinism is usually controlled by a single autosomal gene with two alleles. The allele for normal pigmentation (let's call it R) is dominant to the allele for albinism (r). Individuals with either an RR or Rr genotype will not have albinism, because the R allele is dominant over the recessive r allele.

However, consider what happens if two individuals with the Rr genotype reproduce with each other. The outcome would be similar to the example shown in the Punnett square above for two hypothetical Aa individuals. Their possible offspring could be RR (normal pigmentation), Rr (normal pigmentation), or rr (albinism). This explains why a child with albinism (rr), can have two parents that do not have albinism. Both unaffected parents must be a carrier of the recessive r allele, but they also have a dominant R allele that prevents them from having the condition themselves.

Some other human traits that have a Mendelian inheritance pattern are Huntington's disease and wet versus dry ear wax. You may have heard about other human traits that were *previously* thought to be Mendelian, such as dimples, a widow's peak hairline, hitchiker's thumb, and the ability to roll your tongue. As science has progressed, it is now understood that these are not actually Mendelian traits. In fact, most human traits are actually controlled by multiple genes, or otherwise have more than two alleles, which means they do not have a simple Mendelian inheritance pattern.

X-Linked Mendelian Traits in Humans

Because males have just one X chromosome, they have only one allele for any X-linked trait. Therefore, a recessive X-linked allele is always expressed in males. Because females have two X chromosomes, they have two alleles for any X-linked trait. Therefore, they must inherit two copies of the recessive allele to express an X-linked recessive

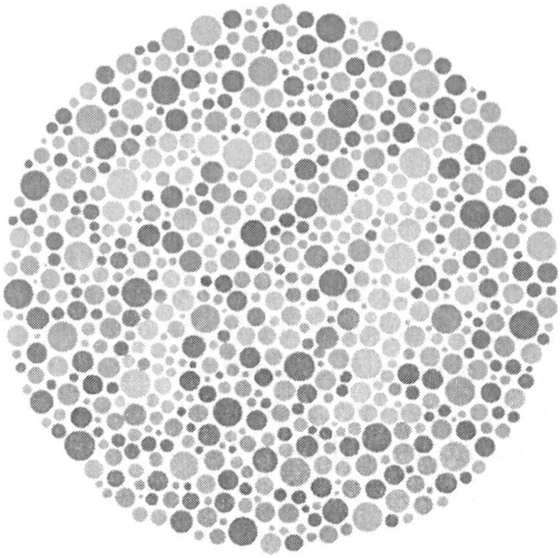

Figure : This circle of colours containing the number 74 is part of the Ishihara colour blindness test.

trait. This explains why X-linked recessive traits are less common in females than males, and why they show a different pattern of inheritance than autosomal traits.

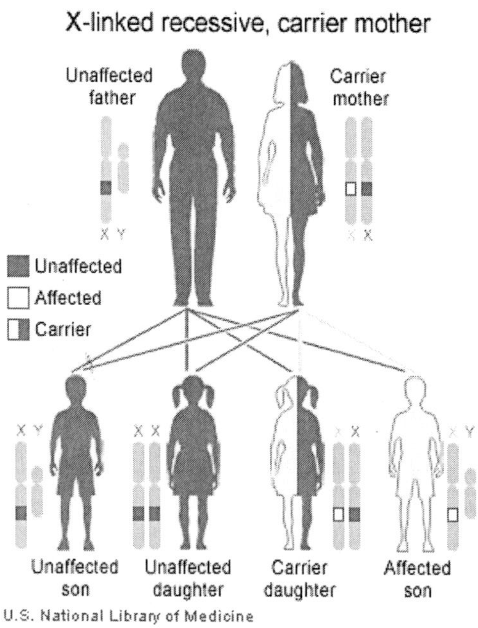

Figure : Heredity implications of an X-linked recessive gene carried by the mother.

An example of a recessive X-linked trait is red-green colour blindness. People with this trait cannot distinguish between the colours red and green. More than one recessive gene on the X chromosome codes for this trait, which is fairly common in males, but relatively rare in females. A female with one of the recessive alleles for the trait does not have the trait herself, but can pass it on to her children. In this case, she is called a carrier of the trait. Half of any sons she has can be expected to have the trait, because there is a 50 per cent chance that they will inherit the X chromosome with the colour-blindness allele. Having only one X chromosome, the recessive allele will be expressed in the sons who inherit it. However, as long as the father is not affected, none of the woman's daughters will have the trait. The daughters have a 50 per cent chance of inheriting the X chromosome with the colour-blindness allele, but it won't be expressed because it is recessive to the normal allele they inherit from their father.

Another example of a recessive X-linked Mendelian trait is hemophilia, which is a disorder characterized by the blood's inability to clot normally. England's Queen Victoria carried the disorder. Two of her five daughters inherited the hemophilia allele from their mother and were carriers. When they married royalty in other European countries, they spread the allele across Europe, including to the royal families of Spain, Germany, and Russia. Victoria's son Prince Leopold also inherited the hemophilia allele from his mother, and he actually suffered from the disease. Understandably, hemophilia was once popularly called "the royal disease."

Feature: My Human Body.

Are you colour blind, or do you think you might be? If you inherited this X-linked recessive disorder, a world without clear differences between certain colours seems normal to you. It's all you have ever known! Some people who are colour blind are not even aware of it. Simple tests have been devised to determine whether a person is colour blind, and to what degree. An example of such a test. What do you see when you look at this circle? Can you clearly perceive the number 74? If so, you probably have normal red-green colour vision. If you cannot see the number, you may have red-green colour blindness.

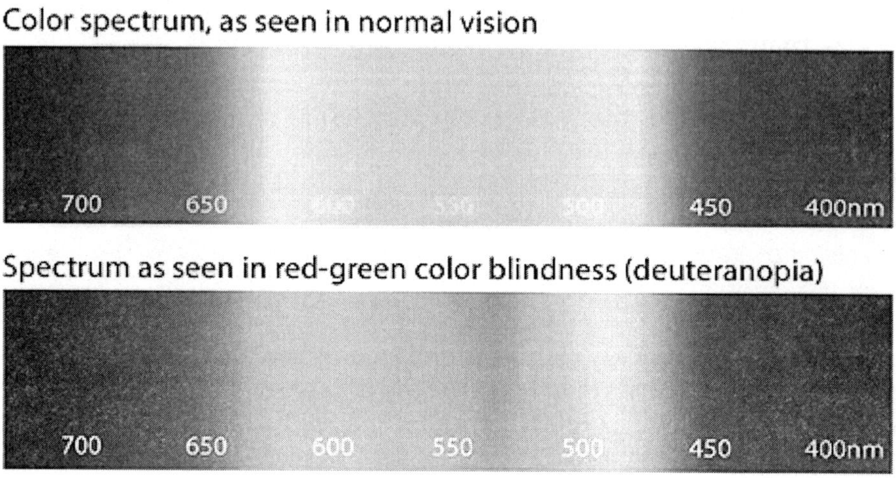

Figure : Perception of colours comparison.

Being colour blind can cause a number of problems. These may range from minor frustrations to outright dangers. Here are a few examples:

- If you are colour blind, it may be difficult to colour-coordinate clothing and furnishings. You may end up wearing colour combinations that people with normal colour vision think are odd or clashing.
- Many LED indicator lights are red or green. Power strips and electronic devices, for example, may have indicator lights to show whether they are on (green) or off (red).
- Test strips for pH, hard water, swimming pool chemicals, and other common tests are also often colour coded. Litmus paper for testing pH, for example, turns red in the presence of an acid, but if you are colour blind, you may not be able to read the test result.
- Do you like your steak well done? If you are colour blind, you may not be able to tell if the meat is still undercooked (red) or grilled just right. You also may not be able to distinguish ripe (red) from unripe (green) fruits and vegetables, such as tomatoes. And some foods, such as dark green spinach, may look more like mud than food, making them totally unappetizing.
- Weather maps often are colour coded. Is that rain (green) in your forecast, or a wintry mix of sleet and freezing rain (pink

or red)? If you can't tell the difference, you may go out on the roads when you shouldn't, putting yourself in danger.

- Being able to distinguish red from green traffic lights may be a matter of life or death. This can be very difficult for someone with red-green colour blindness. In some countries, people with this vision defect are not allowed to drive.

Chapter 3

Interaction of Genes

Gene interaction is the process by which the expression of two or more genes influences one another in different ways as an organism develops a single characteristic. The majority of the traits that comprise living beings are coordinated by various genes.

Mendel and other researchers assumed that characters were controlled by a single gene, but it was later found that multiple characters were controlled by two or more genes. Such genes modify the conventional dihybrid (9:3:3:1) or trihybrid (27:9:9:9:3:3:3:1) ratios by influencing the development of the concerned traits in various ways.

The expression of one gene is influenced by the expression, present or absent, of another gene in a gene interaction.

Types of Gene Interaction

Gene interactions are divided into two categories:

1. **Allelic or Non-epistatic Gene Interaction:** This gene interaction occurs between the alleles of a single gene.
2. **Non-allelic or Epistatic Gene Interaction:** This gene interaction involves interactions between genes on identical or distinct chromosomes.

Allelic or Non-epistatic Gene Interaction

When phenotypic ratios diverge from Mendelian ratios, it is difficult for Mendelian genetics to explain some types of inheritance. This is because specific alleles can often be equally or partially dominant to each other or due to the lethal alleles. Allelic, non-epistatic or intra-allelic interactions are the terms used to describe genetic interactions between alleles of a single gene.

Incomplete Dominance (1:2:1)

A dominant allele could not entirely suppress the other allele. Thus, a heterozygote is phenotypically differentiated from either homozygote (intermediate phenotype).

In snapdragon and *Mirabilis jalapa*, the hybridisation of pure-bred red-flowered and white-flowered plants results in pink-flowered F_1 seedlings (deviated from parental phenotypes), which are intermediate between the two parents. When the F_1 plant self-fertilises, the F_2 progeny displays three classes of plants in the ratio of 1 red:2 pink:1 white rather than 3:1.

Codominance

Here, heterozygotes exhibit the expression of both alleles of a gene. Instead of the intermediate phenotype, the F_1 hybrid displays the phenotypes of both parents. In humans, a single gene regulates the MN blood group.

There are just two alleles: M and N. Children born to a father with the N blood type (genotype NN) and a mother with the M blood type (genotype MM) would have the MN blood type (genotype MN). The hybrid exhibits both phenotypes. Thus, the blood groups that F_2 segregates into are 1M:2MN:1N.

Over Dominance

F_1 heterozygotes occasionally have more extreme phenotypes than either of their parents. The heterozygous white eyes of *Drosophila* have more fluorescent eye pigment than either of its parents.

Lethal Factor

Lethal factors are genes that lead to the death of the person who has them. The heterozygotes are unaffected by recessive lethal because they only express when they are homozygous. Some genes dominantly affect phenotype yet are lethal when expressed in a recessive manner, such as the mouse gene for the yellow coat colour.

Multiple Alleles

There may be more than two allelomorphs or alleles at the same chromosomal location for a gene encoding a specific characteristic (only two are present in a diploid organism). These allelomorphs produce multiple alleles. The best example is the human ABO blood type.

Non-allelic or Epistatic Gene Interaction

When two or more genes affect each other's expression in various ways, non-allelic or epistatic or inter-allelic interaction takes place, resulting in the development of a single character.

Simple Interaction (9:3:3:1)

Here, the same character is influenced by two non-allelic gene pairs. When any of the two factors is present independently, its dominant allele results in different phenotypes. When both dominant alleles are present, a distinctive phenotype is produced. Another phenotype results from the absence of both dominant alleles.

Complementary Factor (9:7)

The interaction of two or more genes inhabiting distinct loci acquired from different parents results in the development of specific characteristics. These genes are complementary to each other, which means they do not express themselves when present separately. Instead, they only express themselves when introduced together through a suitable crossing.

More than two complementary genes can be involved; for example, three complementary genes control the aleurone colour of maize.

Epistasis

Epistasis occurs when a gene or gene pair suppresses or hinders the expression of another non-allelic gene. The gene that causes the effect is referred to as an epistatic gene, while the gene whose expression is inhibited is referred to as a hypostatic gene.

Classification of Epistasis Gene Interaction

Epistatic gene interaction is categorised as follows based on how the involved genes affect one another's expression:

- Supplementary gene interaction
- Complementary gene interaction
- Inhibitory gene interaction
- Duplicate gene interaction
- Masking gene interaction
- Polymeric gene interaction

Supplementary Gene Interaction

In supplemental gene interaction, the phenotypic effect is produced by the dominant allele of one of the two genes regulating a character.

However, the dominant allele of another gene has no independent phenotypic effect.

Therefore, it changes the phenotypic effect brought on by the first gene when it coexists with the dominant allele. For example, Agouti (grey) coat development in mice.

Parents	**CCaa** Coloured ↓		**ccAA** Albino ↓
Gametes	**Ca**	x ↓	**cA**
F_1		**CcAa** ↓	
Selfing of F_1		Agouti	

F_2 Generation:

	CA	**Ca**	**cA**	**ca**
CA	CCAA (Agouti)	CCAa (Agouti)	CcAA (Agouti)	CcAa (Agouti)
Ca	CCAa (Agouti)	CCaa (Coloured)	CcAa (Agouti)	Ccaa (Coloured)
cA	CcAA (Agouti)	CcAa (Agouti)	ccAA (Albino)	ccAa (Albino)
ca	CcAa (Agouti)	Ccaa (Coloured)	ccAa (Albino)	ccaa (Albino)

Thus, the phenotypic ratio is 9 (Agouti):3 (coloured):4 (Albino).

Complementary Gene Interaction

The F_2 ratio becomes 9:7 instead of 9:3:3:1 if both gene loci are homozygous and both produce the same phenotype. In this instance, a single phenotype is produced by the genotype aaBB, aaBb, Aabb, aabb.

When both dominant alleles are present simultaneously, they are referred to as complementary genes and result in a distinct phenotype.

In sweet pea, the homozygous presence of the genes CC, cc, PP, and PP results in no colour (white) as chromogen production is not possible in a homozygous state, but is possible in a heterozygous gene condition.

Inhibitory Gene Interaction

When the heterozygous (Bb) and homozygous (BB) forms of a dominant allele at a single gene locus (B) result in the same phenotype, the F_2 ratio changes from 9:3:3:1 to 13:3. A distinctive phenotype is produced by the homozygous recessive (bb) condition.

It is known as inhibitory gene interaction when a homozygous recessive (bb) trait inhibits the phenotypic expression of the other genes.

Duplicate Gene Interaction

When an identical phenotype is produced by the dominant allele of both gene loci without any cumulative impact, it is referred to as duplicate gene interaction. In this type of interaction, the ratio changes to 15:1 rather than 9:3:3:1. Shepherd's purse plant exhibits duplicate gene interaction.

Plant seed capsules from the shepherd's purse species can be either triangular or oval. When both the genes are present in a homozygous recessive state, the seed capsule adopts an ovoid shape.

Masking Gene Interaction

When the dominant allele (A) of one gene masks the activity of the allele of another gene (B), it is referred to as masking gene interaction. It is thus said that the locus of gene A is epistatic to the locus of gene B.

Dominant epistatic relationships are those in which the dominant allele A expresses itself only when B or b is present. Only when the epistatic locus allele is homozygous recessive does the hypostatic locus allele express itself.

Polymeric Gene Interaction

Two dominant alleles working together to intensify the phenotype or produce a median variance is known as polymeric gene interaction. Each dominant allele alone results in a physical characteristic which is distinct from the combination of dominant alleles. This results in three phenotypes being produced from just two dominant alleles. As a result, neither dominant allele is outperforming the other dominant allele.

Biological Significance

Genetic interactions have significant effects on how genotype and phenotype are related. They have been put out as an explanation for missing heritability. The term "missing heritability" describes how many heritable characteristics still have unidentified genetic ancestors.

Genetic interactions could significantly reduce the amount of missing heritability, despite the many hypotheses that have been presented. These genetic interactions would probably transcend the pairwise interactions taken into account in genetic interaction networks.

Therefore, the expression of genes is not independent of one another and depends on the presence or absence of other genes. This type of deviation from the Mendelian principle of one gene-one trait is termed the Factor Hypothesis or Gene Interaction.

Introduction, interaction involving two gene pairs

Epistatic Relationships Involving Two Genes

As previously mentioned, scientists have performed numerous studies in an attempt to better understand and classify digenic epistatic relationships. Some of the most famous examples of research in which the interaction between two genes was found to produce a novel phenotype are examined in the following sections.

Combs in Chick

Figure : Rooster displaying a single comb.

In the first decade of the twentieth century, British geneticists William Bateson and R. C. Punnett conducted research showing that the shape of the comb in chickens was caused by the interaction between two different genes. Bateson and Punnett were aware of the fact that different varieties of chickens possess distinctive combs. For instance, Wyandottes have a "rose" comb, Brahmas have a "pea" comb, and Leghorns have a "single" comb. When Bateson and Punnett crossed a Wynadotte chicken with a Brahma chicken, all of the F_1 progeny had a new type of comb, which the duo termed a "walnut" comb. In this case, neither the rose comb of the Wyandotte nor the pea comb of the Brahma appeared to be dominant, because the F_1 offspring had their own unique phenotype. Moreover, when two of these F_1 progeny were crossed with each other, some of the members of the resulting F_2 generation had walnut combs, some had rose combs, some had pea combs, and some had a single comb, like that seen in Leghorns. Because the four comb shapes appeared in a 9:3:3:1 ratio (i.e., nine walnut chickens per every three rose chickens per every three pea chickens per every one single-comb chicken), it seemed that two different genes must play a role in comb shape.

Through continued research, Bateson and Punnett deduced that Wyandotte (rose-combed) chickens must have the genotype *RRpp*, while Brahma chickens must have the genotype *rrPP*. A cross between a Wyandotte and a Brahma would yield offspring that all had the *RrPp* genotype, which manifested as the walnut-comb phenotype. Indeed, any chicken with at least one rose-comb allele (*R*) and one pea-comb allele (*P*) would have a walnut comb. Thus, when two F_1 walnut chickens were crossed, the resulting F_2 generation would yield rose-comb chickens (*R_pp*), pea-comb chickens (*rrP_*), and walnut-comb chickens (*R_P_*), as well as chickens with a new, fourth phenotype—the single-comb phenotype. Based on the process of elimination, it could be assumed that these single-comb chickens had the *rrpp* genotype (Bateson & Punnett, 1905; 1906; 1908).

Flower Color in Peas

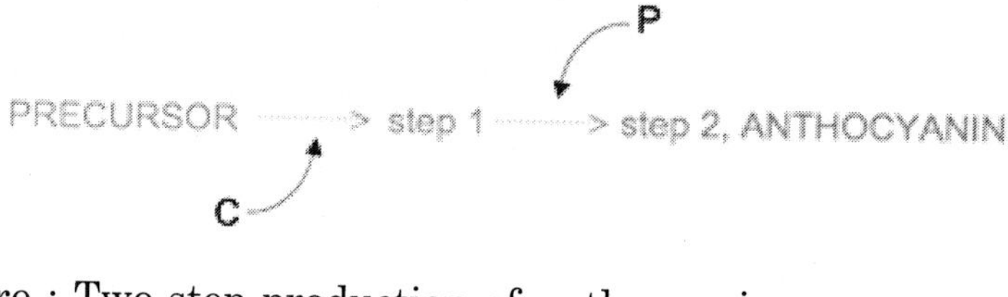

Figure : Two-step production of anthocyanin.

Two genes are responsible for the chemical reaction that produces the plant pigment anthocyanin from a precursor molecule. Gene C controls the first step in the reaction to produce the step 1 product, and gene P controls the second step in the reaction to produce anthocyanin.

Bateson and Punnett next performed a set of experiments in peas that also showed gene-gene interaction. The duo opted to use peas because it is relatively easy to perform hybrid crosses with these plants, and they chose to focus on the characteristic of flower color.

Bateson and Punnett began by crossing two varieties of pea, each of which was pure-breeding for white flowers. This cross yielded an F_1 generation in which all progeny had purple flowers. Next, two F_1 plants were crossed to create the F_2 generation. In this generation, Bateson and Punnett counted a total of 382 purple-flowered plants and 269 white-flowered plants. The ratio of purple flowers to white flowers was thus 9.4:6.6, or approximately 9:7.

What could explain this variation from Mendelian ratios? Bateson set out to answer this question in a 1909 report, in which he first proposed what he called the ability of one "allelomorphic pair" (pair of gene alleles) to mask the affects of the alleles for another gene. To rephrase this in terms of Bateson and Punnett's pea experiment, it seemed that two recessive alleles at one flower locus could mask the effects of the alleles at the other flower locus. Let's designate the first locus as the *C* locus, and the second as the *P* locus. If Bateson's theory held true, it meant that any flower with the *cc* genotype would be white, no matter what alleles were present at its *P* locus. Similarly, any flower with the *pp* genotype would also be white, no matter what alleles were present at its *C* locus. Bateson later used the word "epistasis," which translates as "standing upon," to define the masking action of one gene by another. (Since then, scientists have come to understand that genes can interact in more ways than just masking.)

Many years after Bateson first described this 9:7 phenotypic ratio in pea plants, researchers were finally able to determine the two genes responsible for it (Dooner *et al.*, 1991). These genes control flower color by controlling pea plant biochemistry, in particular that related to pigment compounds called anthocyanins. In peas, there is a two-step chemical reaction that forms anthocyanins; gene *C* is responsible for the first step, and gene *P* is responsible for the second. If either step is nonfunctional, then no purple pigment is produced, and the affected pea plant bears only white flowers. The dominant *C* and *P* alleles code for functional steps in anthocyanin production, whereas the recessive *c* and *p* alleles code for nonfunctional steps. Thus, if two recessive alleles occur for either gene, white flowers will result.

A modification of phenotypic but not genotypic Mendelian ratios. Note that the *C* and *P* genes independently assort, and remember that the presence of a recessive genotype at one locus (i.e., *cc* or *pp*) masks the effects of the alleles at the other locus. Note also that there are nine combinations of alleles in the F_1 generation that feature at least one dominant *C* and one dominant *P* allele, which would yield a purple flower phenotype (indicated within the table by purple shading). Conversely, there are seven combinations that result in either a *cc* or a *pp*, which would yield the white flower phenotype-hence, the 9:7 ratio of purple to white flowers.

Table : Results of the Cross Between Two Pea Plants with Genotype CcPp

		Female Gametes			
		CP	***Cp***	***cP***	***Cp***
Male Gametes	***CP***	*CCPP*	*CCPp*	*CcPP*	*CcPp*
	Cp	*CCPp*	*CCpp*	*CcPp*	*Ccpp*
	cP	*CcPP*	*CcPp*	*ccPP*	*ccPp*
	cp	*CcPp*	*Ccpp*	*ccPp*	*Ccpp*

Primula Petal Color

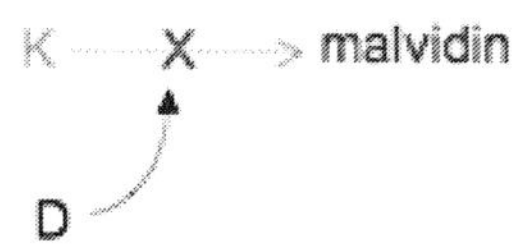

Figure : Malvidin production.

Production of the petal pigment malvidin is controlled by one gene, but its synthesis can be suppressed by another gene at a different locus.

Later researchers discovered that flower petal color can also be controlled by dominant epistatic forces. For instance, in the *Primula* plant, the pigment malvidin creates blue-colored flowers. Synthesis of malvidin is controlled by gene *K*, yet production of this pigment can be suppressed by gene *D*, which is found at completely different locus. In this case, the *D* allele is dominant to the *K* allele, so plants with the genotype *KkDd* will not produce malvidin because of the presence of the *D* allele.

So, if two plants with genotype *KkDd* are crossed with each other, what is the ratio of blue offspring to nonblue offspring?

Table : Results of the Cross Between Two Primula Plants with Genotype KdDd

		Female Gametes			
		KD	***Kd***	***kD***	***kd***
Male Gametes	***KD***	*KKDD*	*KKDd*	*KkDD*	*KkDd*
	Kd	*KKDd*	*KKdd*	*KkDd*	*Kkdd*
	kD	*KkDD*	*KkDd*	*kkDD*	*kkDd*
	kd	*KkDd*	*Kkdd*	*kkDd*	*kkdd*

All of the shaded boxes contain genotypes that indicate an absence of malvidin. In particular, the yellow boxes represent genotypes that feature at least one *D* allele, and the presence of the *D* allele suppresses the production of malvidin. Meanwhile, the light turquoise box represents a genotype with no malvidin suppression (*dd*), but also no malvidin production (*kk*). On the other hand, the three unshaded/white boxes are the only genotypes that permit the production of malvidin, which means that plants with these phenotypes bear blue flowers. The phenotypic ratio is therefore 13:3. This type of epistasis is sometimes called dominant suppression, because the deviation from 9:3:3:1 is caused by a single allele that produces a dominant phenotype, and the action of this allele is to suppress the expression of some other gene.

Wheat Kernel Color

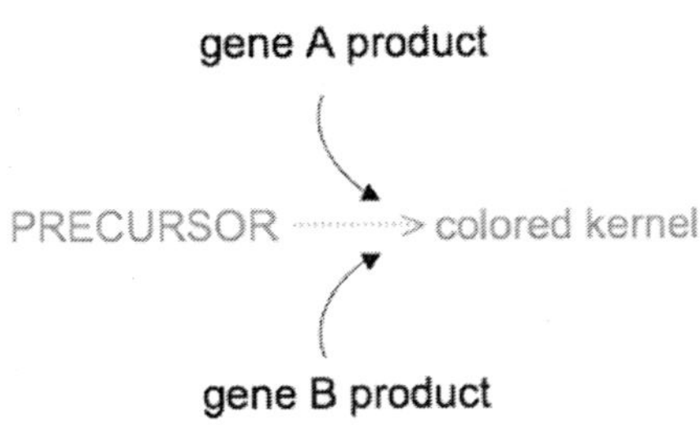

Figure 4: Colored kernel production.

Wheat kernel color is determined by the action of two genes: gene A and gene B. If either gene is functional, a colored kernel will result. If both genes are not functional, the kernel will be colorless.

Genes don't always have to act in opposition to each other for an interaction to be epistatic, however. Sometimes, two genes that each have the same role in protein production can substitute for each other.

The mechanism by which wheat kernel color is determined is an example of duplicate gene action. In wheat, kernel color is dependent upon a biochemical reaction that converts a precursor substance into a pigment, and this reaction can be performed with the product of either gene *A* or gene *B*. Thus, having either an *A* allele or a *B* allele produces color in the kernel, but a lack of either allele will produce a white kernel that is devoid of color.

The results of a dihybrid cross between two plants with genotype *AaBb*. Note that among the offspring of this cross, the phenotype is nearly uniform.

Table : Results of the Cross Between Two Wheat Plants with Genotype AaBb

		Female Gametes			
		AB	***Ab***	***aB***	***ab***
Male Gametes	***AB***	*AABB*	*AABb*	*AaBB*	*AaBb*
	Ab	*AABb*	*AAbb*	*AaBb*	*Aabb*
	aB	*AaBB*	*AaBb*	*aaBB*	*aaBb*
	ab	*AaBb*	*Aabb*	*aaBb*	*aabb*

In this cross, whenever a dominant allele is present at either locus, the biochemical conversion occurs, and a colored kernel results. Thus, only the double homozygous recessive genotype produces a phenotype with no color, and the resulting phenotypic ratio of color to noncolor is 15:1.

The gene *D* is epistatic to gene *K* because one allelic pair masks the expression of an allele at the second locus. In the case of duplicate gene action, the outcome is less variable, but it is still derived from multiple gene interactions. Here, if wheat kernel color is controlled by genes *A* and *B*, then *A* is epistatic to allele *B* or allele *b*, and *B* is epistatic to allele *A* or allele *a*.

Coat Color in Horses

Yet another type of epistasis occurs when one gene interacts with another to modify—but not mask—a phenotype. For example, in horses, the extension gene determines whether an animal's coat color will be red or black; here, the dominant allele *E* produces black pigment in the coat, while the recessive allele *e* produces red pigment. All horses with genotype *ee* are therefore red, yet there are many different types of red horses. These differences exist because of the action of epistatic modifier genes.

One such modifier gene is called cream dilution. The cream dilution gene has two alleles: C^{Cr} and C. The C^{Cr} allele is semidominant; it dilutes red to yellow in the heterozygous state and red to pale cream in the homozygous state. On the other hand, the C allele has no diluting effect on coat color. Thus, horses with genotype *eeCC* are chestnut colored, and they have reddish-brown coats, tails, and manes. In contrast, horses with one copy of the C^{CR} allele (genotype $eeCC^{CR}$) are palomino (i.e., they have a gold coat with a white mane and tail), while horses with two copies of the C^{CR} allele (genotype $ee\ C^{CR}C^{CR}$) are cremello (i.e., basically white or cream colored).

Interaction involving more than two gene pairs

Most heritable traits are affected by inheritance of alleles at multiple loci, and the identification of these loci is a key challenge of modern genetic research. We recently showed that gene expression levels in yeast typically show multigenic inheritance and provide a good model for investigating the genetic basis of complex traits. Here we use this system to examine the prevalence of genetic interactions in a large set of phenotypes. A genetic interaction between a pair of loci (sometimes termed epistasis) occurs when the effect of an allele at one locus changes as a function of the allele at the other. Previous biometric analyses have provided evidence for many interactions underlying transcript levels. We sought to identify the loci involved in such interactions in a cross between a lab strain of yeast, BY, and a wild strain, RM. For this analysis, we used previously described genotype and gene expression data from 112 segregants. We first tested all possible pairs of loci for evidence of interaction underlying each transcript level. This strategy had little statistical power to map interactions due to its requirement for a very large number of tests and the corresponding stringent correction of significance thresholds for multiple testing.

To improve power, we used a two-stage search strategy. For each transcript, we first identified the "primary" QTL with the strongest individual linkage. We then partitioned segregants based on inheritance (either BY or RM) at the primary locus and tested each subgroup for further "secondary" loci. We computed the joint significance of the strongest such secondary QTL with the primary locus using a newly developed statistical method based on estimation of the false discovery rate. This method provides an estimate of the overall fraction of transcripts for which both loci are involved in the genetics, as well as a probability for each transcript that both loci are true positives. This analysis indicated that the locus pairs identified for 57% of all transcripts were true positives, and it identified locus pairs for 225 transcripts at a false discovery rate of 5% (*i.e.* fewer than 12 of these are expected to have either locus as a false positive).

Because the two-stage search takes account of inheritance at the primary locus in assessing the effect of the secondary locus, it can identify locus pairs with interactions, as well as locus pairs that act additively. We tested each of the 225 locus pairs for interacting effects by fitting a regression model relating inheritance at the linking loci to the transcript level. The model included an additive effect for each locus and a term representing the interaction between loci; the latter term was significant at $p < 0.05$ for 65% of transcripts. For comparison, we considered 547 transcripts for which two loci were identified as significant by their individual effects. Of these, only 13% had an interaction term significant at $p < 0.05$. A false discovery rate analysis indicated that the interaction term represented a true positive for 91% of locus pairs mapped by the two-stage search, compared to 13% of pairs mapped independently.

We next sought to determine whether the secondary loci mapped by the two-stage search could have been detected by a single-locus linkage test. Under some interaction scenarios, the individual effects of the loci are undetectable, while under others the loci retain an individual effect. We tested the individual effect of each secondary locus on the corresponding transcript, and found that 87% of these were significant at $p < 0.05$, indicating that most secondary loci do have a residual individual effect. However, this significance threshold does not take into account the multiple tests carried out in a genome scan. When we asked how many of the secondary loci were detectable at a false discovery rate of 5% in the context of a genome scan, we found that only 74 (33%) of the 225 were, indicating that 67% of secondary loci would have been missed without the two-stage search.

Pleiotropic genes penetrance and expressivity

Pleiotropy

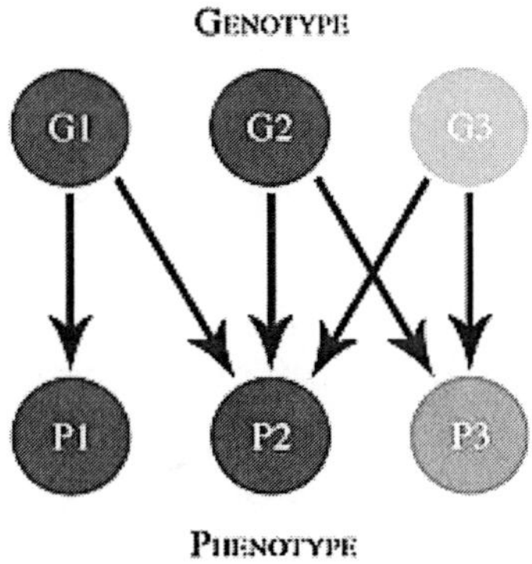

Simple genotype–phenotype map that only shows additive pleiotropy effects. G1, G2, and G3 are different genes that contribute to phenotypic traits P1, P2, and P3.

Pleiotropy (from Greek ðëåßùí *pleion*, 'more', and ôñüðïò *tropos*, 'way') occurs when one gene influences two or more seemingly unrelated phenotypic traits. Such a gene that exhibits multiple phenotypic expression is called a pleiotropic gene. Mutation in a pleiotropic gene may have an effect on several traits simultaneously, due to the gene coding for a product used by a myriad of cells or different targets that have the same signaling function.

Pleiotropy can arise from several distinct but potentially overlapping mechanisms, such as gene pleiotropy, developmental pleiotropy, and selectional pleiotropy. Gene pleiotropy occurs when a gene product interacts with multiple other proteins or catalyzes multiple reactions. Developmental pleiotropy occurs when mutations have multiple effects on the resulting phenotype. Selectional pleiotropy occurs when the resulting phenotype has many effects on fitness (depending on factors such as age and gender).

An example of pleiotropy is phenylketonuria, an inherited disorder that affects the level of phenylalanine, an amino acid that can be obtained from food, in the human body. Phenylketonuria causes this amino acid to increase in amount in the body, which can be very dangerous. The disease is caused by a defect in a single gene on chromosome 12 that codes for enzyme phenylalanine hydroxylase, that affects multiple systems, such as the nervous and integumentary system.

Pleiotropic gene action can limit the rate of multivariate evolution when natural selection, sexual selection or artificial selection on one trait favors one allele, while selection on other traits favors a different allele. Some gene evolution is harmful to an organism. Genetic correlations and responses to selection most often exemplify pleiotropy.

History

Pleiotropic traits had been previously recognized in the scientific community but had not been experimented on until Gregor Mendel's 1866 pea plant experiment. Mendel recognized that certain pea plant traits (seed coat color, flower color, and axial spots) seemed to be inherited together; however, their correlation to a single gene has never been proven. The term "pleiotropie" was first coined by Ludwig Plate in his Festschrift, which was published in 1910. He originally defined pleiotropy as occurring when "several characteristics are dependent upon ... [inheritance]; these characteristics will then always appear together and may thus appear correlated". This definition is still used today.

After Plate's definition, Hans Gruneberg was the first to study the mechanisms of pleiotropy. In 1938 Gruneberg published an article dividing pleiotropy into two distinct types: "genuine" and "spurious" pleiotropy. "Genuine" pleiotropy is when two distinct primary products arise from one locus. "Spurious" pleiotropy, on the other hand, is either when one primary product is utilized in different ways or when one primary product initiates a cascade of events with different phenotypic consequences. Gruneberg came to these distinctions after experimenting on rats with skeletal mutations. He recognized that "spurious" pleiotropy was present in the mutation, while "genuine" pleiotropy was not, thus partially invalidating his own original theory. Through subsequent research, it has been established that Gruneberg's definition of "spurious" pleiotropy is what we now identify simply as "pleiotropy".

In 1941 American geneticists George Beadle and Edward Tatum further invalidated Gruneberg's definition of "genuine" pleiotropy, advocating instead for the "one gene-one enzyme" hypothesis that was originally introduced by French biologist Lucien Cuénot in 1903. This hypothesis shifted future research regarding pleiotropy towards how a single gene can produce various phenotypes.

In the mid-1950s Richard Goldschmidt and Ernst Hadorn, through separate individual research, reinforced the faultiness of "genuine" pleiotropy. A few years later, Hadorn partitioned pleiotropy into a "mosaic" model (which states that one locus directly affects two phenotypic traits) and a "relational" model (which is analogous to "spurious" pleiotropy). These terms are no longer in use but have contributed to the current understanding of pleiotropy.

By accepting the one gene-one enzyme hypothesis, scientists instead focused on how uncoupled phenotypic traits can be affected by genetic recombination and mutations, applying it to populations and evolution. This view of pleiotropy, "universal pleiotropy", defined as locus mutations being capable of affecting essentially all traits, was first implied by Ronald Fisher's Geometric Model in 1930. This mathematical model illustrates how evolutionary fitness depends on the independence of phenotypic variation from random changes (that is, mutations). It theorizes that an increasing phenotypic independence corresponds to a decrease in the likelihood that a given mutation will result in an increase in fitness. Expanding on Fisher's work, Sewall Wright provided more evidence in his 1968 book *Evolution and the Genetics of Populations: Genetic and Biometric Foundations* by using molecular genetics to support the idea of "universal pleiotropy". The concepts of these various studies on evolution have seeded numerous other research projects relating to individual fitness.

In 1957 evolutionary biologist George C. Williams theorized that antagonistic effects will be exhibited during an organism's life cycle if it is closely linked and pleiotropic. Natural selection favors genes that are more beneficial prior to reproduction than after (leading to an increase in reproductive success). Knowing this, Williams argued that if only close linkage was present, then beneficial traits will occur both before and after reproduction due to natural selection. This, however, is not observed in nature, and thus antagonistic pleiotropy contributes to the slow deterioration with age (senescence).

Mechanism

Pleiotropy describes the genetic effect of a single gene on multiple phenotypic traits. The underlying mechanism is genes that code for a product that is either used by various cells or has a cascade-like signaling function that affects various targets.

Polygenic traits

Pleiotropy seems limited for many traits in humans since the SNP overlap, as measured by variance accounted for, between many polygenic predictors is small.

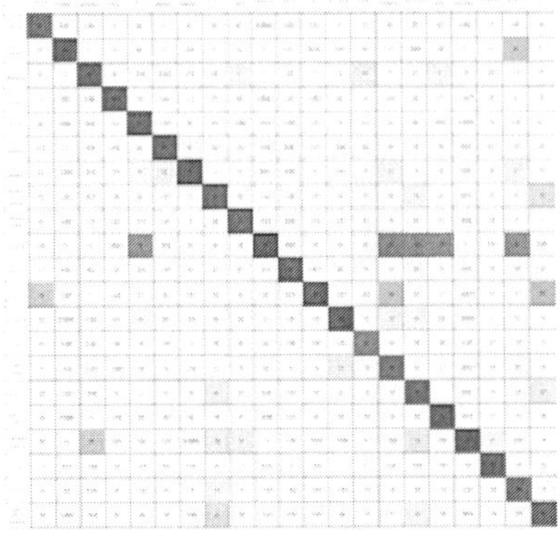

Most genetic traits are polygenic in nature: controlled by many genetic variants, each of small effect. These genetic variants can reside in protein coding or non-coding regions of the genome. In this context pleiotropy refers to the influence that a specific genetic variant, e.g., a single nucleotide polymorphism or SNP, has on two or more distinct traits.

Genome-wide association studies (GWAS) and machine learning analysis of large genomic datasets have led to the construction of SNP based polygenic predictors for human traits such as height, bone density, and many disease risks. Similar predictors exist for plant and animal species and are used in agricultural breeding.

One measure of pleiotropy is the fraction of genetic variance that is common between two distinct complex human traits: e.g., height vs bone density, breast cancer vs heart attack risk, or diabetes vs hypothyroidism risk. This has been calculated for hundreds of pairs of traits, with results. In most cases examined the genomic regions controlling each trait are largely disjoint, with only modest overlap.

Thus, at least for complex human traits so far examined, pleiotropy is limited in extent.

Models for the origin

One basic model of pleiotropy's origin describes a single gene locus to the expression of a certain trait. The locus affects the expressed trait only through changing the expression of other loci. Over time, that locus would affect two traits by interacting with a second locus. Directional selection for both traits during the same time period would increase the positive correlation between the traits, while selection on only one trait would decrease the positive correlation between the two traits.

Eventually, traits that underwent directional selection simultaneously were linked by a single gene, resulting in pleiotropy.

Other more complex models compensate for some of the basic model's oversights, such as multiple traits or assumptions about how the loci affect the traits. They also propose the idea that pleiotropy increases the phenotypic variation of both traits since a single mutation on a gene would have twice the effect.

Evolution

Pleiotropy can have an effect on the evolutionary rate of genes and allele frequencies. Traditionally, models of pleiotropy have predicted that evolutionary rate of genes is related negatively with pleiotropy – as the number of traits of an organism increases, the evolutionary rates of genes in the organism's population decrease. However, this relationship has not been clearly found in empirical studies.

In mating, for many animals the signals and receptors of sexual communication may have evolved simultaneously as the expression of a single gene, instead of the result of selection on two independent genes, one that affects the signaling trait and one that affects the receptor trait. In such a case, pleiotropy would facilitate mating and survival. However, pleiotropy can act negatively as well. A study on seed beetles found that intralocus sexual conflict arises when selection for certain alleles of a gene that are beneficial for one sex causes expression of potentially harmful traits by the same gene in the other sex, especially if the gene is located on an autosomal chromosome.

Pleiotropic genes act as an arbitrating force in speciation. William R. Rice and Ellen E. Hostert (1993) conclude that the observed prezygotic isolation in their studies is a product of pleiotropy's balancing role in indirect selection. By imitating the traits of all-infertile hybridized species, they noticed that the fertilization of eggs was prevented in all eight of their separate studies, a likely effect of pleiotropic genes on speciation. Likewise, pleiotropic gene's stabilizing selection allows for the allele frequency to be altered.

Studies on fungal evolutionary genomics have shown pleiotropic traits that simultaneously affect adaptation and reproductive isolation, converting adaptations directly to speciation. A particularly telling case of this effect is host specificity in pathogenic ascomycetes and specifically, in *venturia*, the fungus responsible for apple scab. These parasitic fungi each adapts to a host, and are only able to mate within a shared host after obtaining resources. Since a single toxin gene or virulence allele can grant the ability to colonize the host, adaptation and reproductive isolation are instantly facilitated, and in turn, pleiotropically causes adaptive speciation. The studies on fungal evolutionary genomics will further elucidate the earliest stages of divergence as a result of gene flow, and provide insight into pleiotropically induced adaptive divergence in other eukaryotes.

Antagonistic pleiotropy

Sometimes, a pleiotropic gene may be both harmful and beneficial to an organism, which is referred to as *antagonistic pleiotropy*. This may occur when the trait is beneficial for the organism's early life, but not its late life. Such "trade-offs" are possible since natural selection affects traits expressed earlier in life, when most organisms are most fertile, more than traits expressed later in life.

This idea is central to the antagonistic pleiotropy hypothesis, which was first developed by G. C. Williams in 1957. Williams suggested that some genes responsible for increased fitness in the younger, fertile organism contribute to decreased fitness later in life, which may give an evolutionary explanation for senescence. An example is the p53 gene, which suppresses cancer but also suppresses stem cells, which replenish worn-out tissue.

Unfortunately, the process of antagonistic pleiotropy may result in an altered evolutionary path with delayed adaptation, in addition to effectively cutting the overall benefit of any alleles by roughly half. However, antagonistic pleiotropy also lends greater evolutionary "staying power" to genes controlling beneficial traits, since an organism with a mutation to those genes would have a decreased chance of successfully reproducing, as multiple traits would be affected, potentially for the worse.

Sickle cell anemia is a classic example of the mixed benefit given by the staying power of pleiotropic genes, as the mutation to Hb-S provides the fitness benefit of malaria resistance to heterozygotes, while homozygotes have significantly lowered life expectancy. Since both of these states are linked to the same mutated gene, large populations today are susceptible to sickle cell despite it being a fitness-impairing genetic disorder.

Examples

Peacock with albinism

Albinism

Albinism is the mutation of the TYR gene, also termed tyrosinase. This mutation causes the most common form of albinism. The mutation alters the production of melanin, thereby affecting melanin-related and other dependent traits throughout the organism. Melanin is a substance made by the body that is used to absorb light and provides coloration to the skin. Indications of albinism are the absence of color in an organism's eyes, hair, and skin, due to the lack of melanin. Some forms of albinism are also known to have symptoms that manifest themselves through rapid-eye movement, light sensitivity, and strabismus.

Autism and schizophrenia

Pleiotropy in genes has been linked between certain psychiatric disorders as well. Deletion in the 22q11.2 region of chromosome 22 has been associated with schizophrenia and autism. Schizophrenia and autism are linked to the same gene deletion but manifest very differently from each other. The resulting phenotype depends on the stage of life at which the individual develops the disorder. Childhood manifestation of the gene deletion is typically associated with autism, while adolescent and later expression of the gene deletion often manifests in schizophrenia or other psychotic disorders. Though the disorders are linked by genetics, there is no increased risk found for adult schizophrenia in patients who experienced autism in childhood.

A 2013 study also genetically linked five psychiatric disorders, including schizophrenia and autism. The link was a single nucleotide polymorphism of two genes involved in calcium channel signaling with neurons. One of these genes, CACNA1C, has been found to influence cognition. It has been associated with autism, as well as linked in studies to schizophrenia and bipolar disorder. These particular studies show clustering of these diseases within patients themselves or families. The estimated heritability of schizophrenia is 70% to 90%, therefore the pleiotropy of genes is crucial since it causes an increased risk for certain psychotic disorders and can aid psychiatric diagnosis.

Phenylketonuria (PKU)

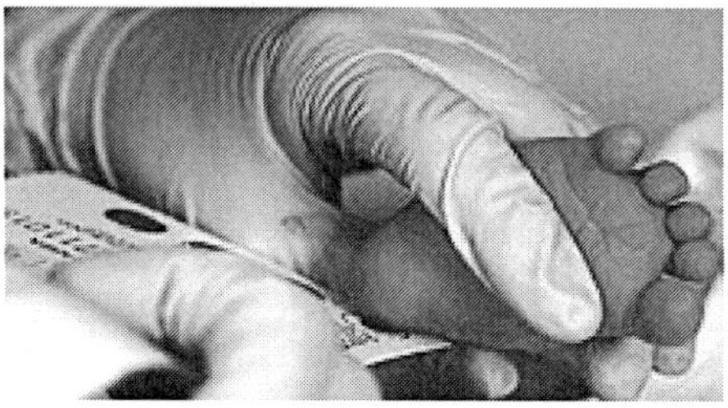

The blood of a two-week-old infant is collected for a PKU screening.

A common example of pleiotropy is the human disease phenylketonuria (PKU). This disease causes mental retardation and reduced hair and skin pigmentation, and can be caused by any of a large number of mutations in the single gene on chromosome 12 that codes for the enzyme phenylalanine hydroxylase, which converts the amino acid phenylalanine to tyrosine. Depending on the mutation involved, this conversion is reduced or ceases entirely. Unconverted phenylalanine builds up in the bloodstream and can lead to levels that are toxic to the developing nervous system of newborn and infant children. The most dangerous form of this is called classic PKU, which is common in infants. The baby seems normal at first but actually incurs permanent intellectual disability. This can cause symptoms such as mental retardation, abnormal gait and posture, and delayed growth. Because tyrosine is used by the body to make melanin (a component of the pigment found in the hair and skin), failure to convert normal levels of phenylalanine to tyrosine can lead to fair hair and skin. The frequency of this disease varies greatly. Specifically, in the United States, PKU is found at a rate of nearly 1 in 10,000 births. Due to newborn screening, doctors are able to detect PKU in a baby sooner. This allows them to start treatment early, preventing the baby from suffering from the severe effects of PKU. PKU is caused by a mutation in the PAH gene, whose role is to instruct the body on how to make phenylalanine hydroxylase. Phenylalanine hydroxylase is what converts the phenylalanine, taken in through diet, into other things that the body can use. The mutation often decreases the effectiveness or rate at which the hydroxylase breaks down the phenylalanine. This is what causes the phenylalanine to build up in the body. The way to treat PKU is to manage one's diet. Phenylalanine is ingested through food, so a diet should decrease types of foods that have high amounts of phenylalanine. Foods with high levels of protein must be avoided. These include breast milk, eggs, chicken, beef, pork, fish, nuts, and other foods. A special PKU formula can be obtained in order for the body to have protein.

Sickle cell anemia

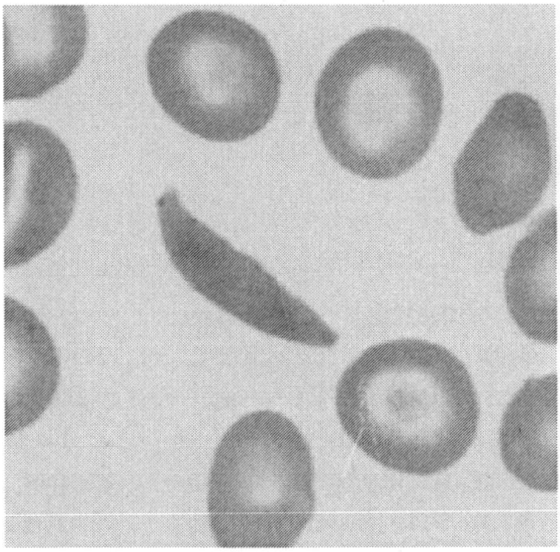

Photomicrograph of normal-shaped and sickle-shape red blood cells from a patient with sickle cell disease

Sickle cell anemia is a genetic disease that causes deformed red blood cells with a rigid, crescent shape instead of the normal flexible, round shape. It is caused by a change in one nucleotide, a point mutation in the HBB gene. The HBB gene encodes information to make the beta-globin subunit of hemoglobin, which is the protein red blood cells use to carry oxygen throughout the body. Sickle cell anemia occurs when the HBB gene mutation causes both beta-globin subunits of hemoglobin to change into hemoglobin S (HbS).

Sickle cell anemia is a pleiotropic disease because the expression of a single mutated HBB gene produces numerous consequences throughout the body. The mutated hemoglobin forms polymers and clumps together causing the deoxygenated sickle red blood cells to assume the disfigured sickle shape. As a result, the cells are inflexible and cannot easily flow through blood vessels, increasing the risk of blood clots and possibly depriving vital organs of oxygen. Some complications associated with sickle cell anemia include pain, damaged organs, strokes, high blood pressure, and loss of vision. Sickle red blood cells also have a shortened lifespan and die prematurely.

Marfan syndrome

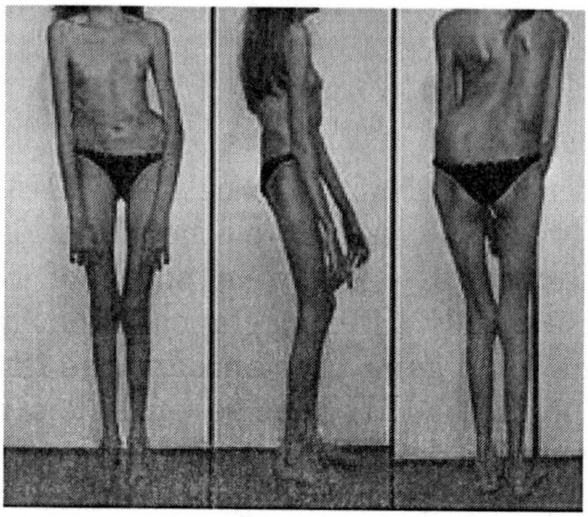

Patient with Marfan Syndrome

Marfan syndrome (MFS) is an autosomal dominant disorder which affects 1 in 5–10,000 people. MFS arises from a mutation in the FBN1 gene, which encodes for the glycoprotein fibrillin-1, a major constituent of extracellular microfibrils which form connective tissues. Over 1,000 different mutations in FBN1 have been found to result in abnormal function of fibrillin, which consequently relates to connective tissues elongating progressively and weakening. Because these fibers are found in tissues throughout the body, mutations in this gene can have a widespread effect on certain systems, including the skeletal, cardiovascular, and nervous system, as well as the eyes and lungs.

Without medical intervention, prognosis of Marfan syndrome can range from moderate to life-threatening, with 90% of known causes of death in diagnosed patients relating to cardiovascular complications and congestive cardiac failure. Other characteristics of MFS include an increased arm span and decreased upper to lower body ratio.

"Mini-muscle" allele

A gene recently discovered in laboratory house mice, termed "mini-muscle", causes, when mutated, a 50% reduction in hindlimb muscle mass as its primary effect (the phenotypic effect by which it was originally identified). In addition to smaller hindlimb muscle mass, the mutant mice exhibit lower heart rates during physical activity, and a higher endurance. Mini Muscle Mice also exhibit larger kidneys and livers. All of these morphological deviations influence the behavior and metabolism of the mouse. For example, mice with the Mini Muscle mutation were observed to have a higher per-gram aerobic capacity.

The mini-muscle allele shows a mendelian recessive behavior. The mutation is a single nucleotide polymorphism (SNP) in an intron of the myosin heavy polypeptide 4 gene.

DNA repair proteins

DNA repair pathways that repair damage to cellular DNA use many different proteins. These proteins often have other functions in addition to DNA repair. In humans, defects in some of these multifunctional proteins can cause widely differing clinical phenotypes. As an example, mutations in the XPB gene that encodes the largest subunit of the basal Transcription factor II H have several pleiotropic effects. XPB mutations are known to be deficient in nucleotide excision repair of DNA and in the quite separate process of gene transcription. In humans, *XPB* mutations can give rise to the cancer-prone disorder xeroderma pigmentosum or the noncancer-prone multisystem disorder trichothiodystrophy. Another example in humans is the *ERCC6* gene, which encodes a protein that mediates DNA repair, transcription, and other cellular processes throughout the body. Mutations in *ERCC6* are associated with disorders of the eye (retinal dystrophy), heart (cardiac arrhythmias), and immune system (lymphocyte immunodeficiency).

Chickens

Chicken exhibiting the frizzle feather trait

Chickens exhibit various traits affected by pleiotropic genes. Some chickens exhibit frizzle feather trait, where their feathers all curl outward and upward rather than lying flat against the body. Frizzle feather was found to stem from a deletion in the genomic region coding for á-Keratin. This gene seems to pleiotropically lead to other abnormalities like increased metabolism, higher food consumption, accelerated heart rate, and delayed sexual maturity.

Domesticated chickens underwent a rapid selection process that led to unrelated phenotypes having high correlations, suggesting pleiotropic, or at least close linkage, effects between comb mass and physiological structures related to reproductive abilities. Both males and females with larger combs have higher bone density and strength, which allows females to deposit more calcium into eggshells. This linkage is further evidenced by the fact that two of the genes, HAO1 and BMP2, affecting medullary bone (the part of the bone that transfers calcium into developing eggshells) are located at the same locus as the gene affecting comb mass. HAO1 and BMP2 also display pleiotropic effects with commonly desired domestic chicken behavior; those chickens who express higher levels of these two genes in bone tissue produce more eggs and display less egg incubation behavior.

Penetrance and expressivity

In this section, we have emphasized that a gene can not determine the phenotype in isolation, but its effect on phenotype is controlled both by other genes and by the environment. These factors controlling the expression of a gene are complex and their precise nature can not be established easily. In order to describe this situation, the two useful terms often used in genetics are **penetrance** and **expressivity.**

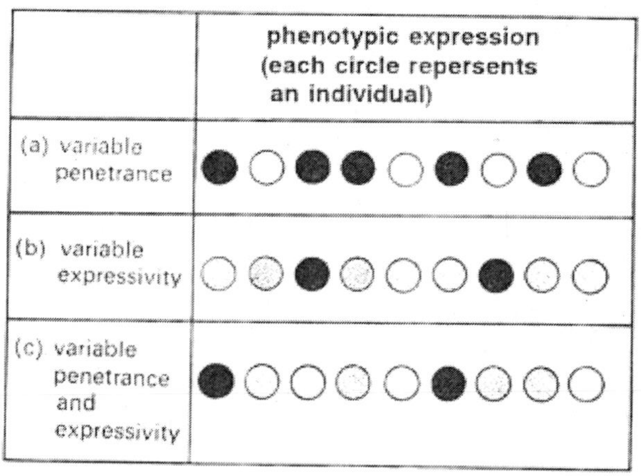

Fig.. Effects of penetrance and expressivity on a hypothetical trait 'pigment intensity'. In each row all individuals have same genotype.

Penetrance is defined as the percentage of individuals with a given genotype, which exhibit the phenotype associated with this genotype. For instance, all individuals with the genotypes *AA* or *Aa,* may not express the dominant phenotype, due to -either the presence of modifiers, suppressors, epistatic genes, etc. or due to the modifying effect of the environment. If only 80% of individuals with *AA* or *Aa* express dominant phenotype, it will be described as 80% penetrance of gene (or allele) *'A'* Expressivity on the other hand, describes the degree or extent to which a given genotype is expressed phenotypically in a particular individual. This may also be due to lack of full expression due to rest of the genome and the environment. For instance, the red eye colour in *Drosophila* flies may show different shades of colour (within the norm of reaction) suggesting variable expression of wild allele *'W'*- in *WW* or *Ww* flies. The distinction between **penetrance** and **expressivity**, and variable expression of coat colour in dogs.

In human pedigree analysis, variable penetrance and expressivity often create problems. For instance, if an allele causing a disease does not have 100% penetrance, an individual's phenotype in the pedigree may be disease free, but the individual may still carry the disease genotype leading to wrong classification.

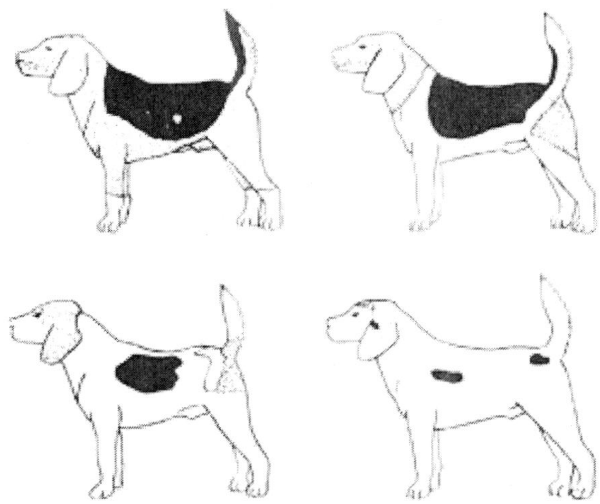

Fig. Four grades of piebald spotting in dogs attributed to variable expressivity.

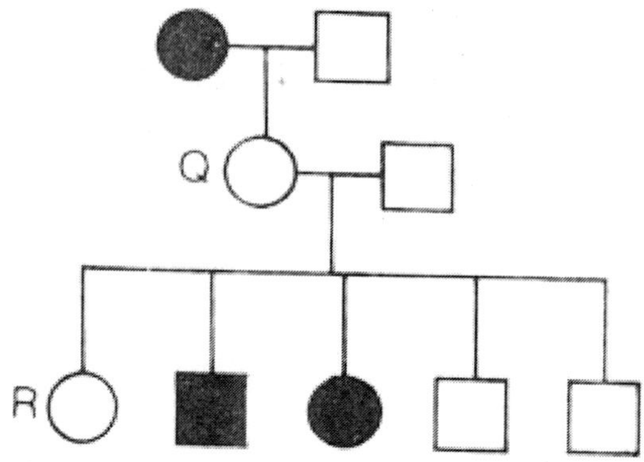

Fig. Lack of penetranee of a dominant gene in an individual Q included fn a pedigree (the gene is passed on to a part of the progeny).

Chapter 4

Multiple Alleles

Alleles or allelomorphs are the alternative forms of a gene present at the same locus on the homologous chromosomes. Some genes have more than two allelic forms, which is referred to as multiple alleles. A diploid cell contains any of the two alleles, one each on the homologous chromosomes and a haploid cell contains only one allele.

When a trait is governed by multiple alleles it is called multiple allelism

Multiple Alleles Characteristics

- Multiple alleles of a gene occupy the same locus on the chromosome.
- Multiple alleles are the alternative forms of the same gene so they influence the same trait.
- The wild-type allele is mostly dominant over the mutant alleles.
- The wild type is considered the standard and all other alleles are considered variants.
- The variant or mutant allele may be dominant, recessive or have an intermediate phenotypic effect.
- Multiple alleles exist at the population level and an individual can have only two alleles for the given gene.
- An individual can have only one type of allele (homozygous) or two different alleles (heterozygous).
- A series of mutant alleles can exist in a population. They are generated by a spontaneous mutation in the wild type and the mutant alleles.
- Multiple alleles lead to the formation of varied phenotypic traits for a character in the population.

Multiple Alleles Examples

The most common examples of multiple alleles are the coat colour of rabbits, A, B, AB and O blood groups in humans and the eye colour in *Drosophila*.

1. **Blood Groups in Humans**

   The A, B, AB and O blood group types in humans are determined by three alleles of a gene. The gene I, which determines the blood type, exists in three allelic forms, which are I^A, I^B and i. The allele I^A produces surface antigen A on the red blood cells and the allele I^B produces the surface antigen B on the red blood cells and the allele i specifies no antigen. Allele I^A and I^B are codominant over allele i.

   A person has any of the two alleles out of the three alleles on the homologous chromosomes. If a person has two I^A alleles (I^AI^A) or one I^A and one i allele (I^Ai) then the blood group is A group. Similarly, I^BI^B and I^Bi give rise to B blood type, I^AI^B gives rise to AB blood type, and two i alleles (ii) give rise to O blood type.

2. **Coat Colour in Rabbit**

   The coat colour of rabbits is determined by the C gene. The C gene has multiple allelic forms that lead to varied phenotypes. The different phenotypes for coat colour in rabbits are agouti (full colour), chinchilla, Himalayan and albino.

   The four common allele of C gene are C, c^{ch}, c^h and c. A rabbit with genotype CC has black or brown fur, $c^{ch}c^{ch}$ have chinchilla or greyish coat colour, c^hc^h have Himalayan pattern and cc have white coat colour, i.e. albino.

   Due to multiple alleles, there are many combinations of alleles possible.

   - The black C allele is dominant over the other three alleles.
   - The c^{ch} (chinchilla) allele shows incomplete dominance over Himalayan and albino alleles.
   - The c^h (Himalayan) allele shows dominance over the albino allele (c).

3. **Eye Colour in Drosophila**

   The eye colour in *Drosophila* is another trait that is determined by multiple alleles. There are a dozen different alleles that show phenotype between the wild-type red colour (w^+) and white colour (w). E.g. coral, blood, eosin, cherry, apricot, etc. The wild-type red colour is dominant over all other mutant alleles and the white colour is recessive to all the alleles.

Classical concept

In classical genetics, a distinction was made between gene and allele on the basis of following two criteria.

Recombination test.

Recombination was believed to take place between two genes but not between two alleles. In other words, intragenic interallelic recombination was not conceived. For instance; a hybrid *aB/Ab* between mutants *aa* (*aaBB*)and *bb* (*AAbb*)for two linked genes *A* and *B* could give rise to wild type progeny on a test cross. Since *A* and *B* are linked, this would be possible only due to recombination. On the other hand if two mutants a_1a_1 and a_2a_2 belonging to same gene *A* were crossed and F_1 $(a_1/a_2)$is test crossed $(a_1a_2 \text{ x } aa)$no wild type progeny would be expected. It was thus, earlier believed that different genes or loci could recombine with each other by crossing over but different alleles of a gene could not.

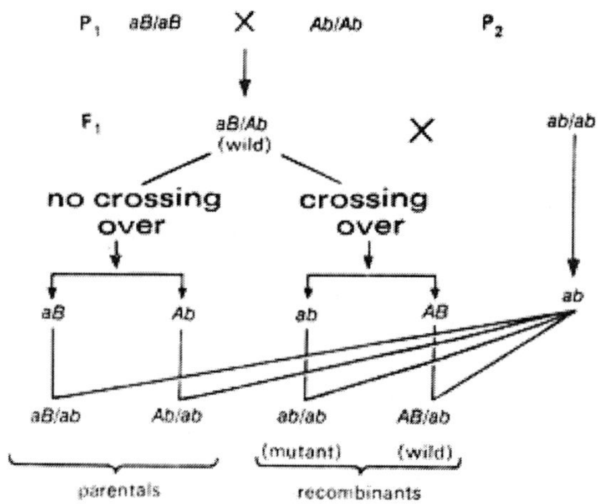

Fig.. Recombination test (a test cross) in which two linked genes in repulsion phase (*aB/Ab*)recombine and give rise to wild type individuals.

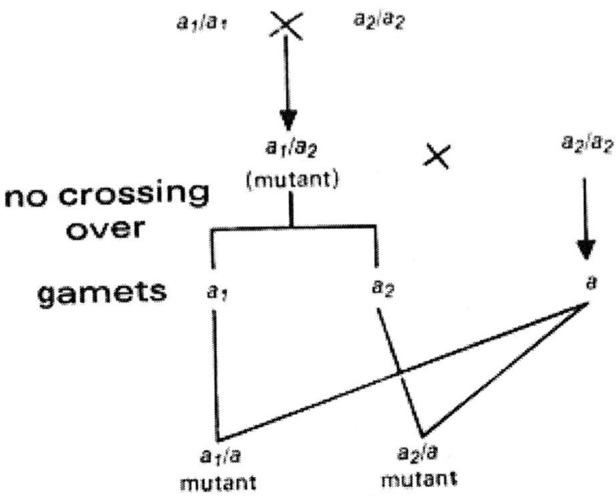

Fig. A cross between two mutants of same gene showing lack of recombination as shown by absence of wild type individuals.

aa × *bb*

↓

F_1 *a*/*b*

F_2 (1) If wild type appears, *a* and *b* are non-allelic.

(2) If no wild type appears, *a* and *b* are allelic.

Fig. A test of allelism (classical concept).

$$\frac{+\;+}{a\;b} \qquad \frac{a\;+}{+\;b}$$

cis (wild) trans (mutant)

Fig. *Cis* and *trans* arrangements of alleles *a* and *b* in heterozygous condition.

Complementation test.

It was also shown that mutant alleles of two different genes coming from two parents, thus being in repulsion phase (also known as *trans* configuration), will complement giving rise to wild type in F_1 generation. But the mutant forms allelic to each other, will never complement.

Thus, in its classical concept, the gene recombined as a unit, functioned as a unit and also changed (mutated) as a unit. This concept was consistent with the view of Morgan and his group that genes were like **beads on a string,** where one bead could change independently and recombine with its neighbouring beads. It will be seen in this section that such a view where genes corresponded to beads on a string was an oversimplification. The first exception to this came, when it was shown that *Bar* locus in *Drosophila* controlling size (number of facets) of eye, contained more than one units of function and could undergo intralocus recombination.

Multiple allele in corn

Several series of multiple alleles are known in corn. Two such series are briefly described.

***C* locus on chromosome 9.** *C* locus on chromosomes 9 controls colour and has atleast three alleles including C^I, an inhibitor of colour, O responsible for full colour in presence of *A* and *R,* and c expressing colourless phenotype. Sometimes they are also designated as *CI, C* and c. These alleles are also described to be closely linked genes, rather than alleles.

***R* locus on chromosome 10.** This is a complex locus, controlling anthocyanin colour of seed and the plant. Following four alleles were earlier designated as R^r, R^g, r^r, and r^g. But later, due to their two independent functions, they were designated as *SP, Sp, sP* and *sp* respectively.

$R^r = SP$ - coloured seed and coloured (red) plant

$R^g = Sp$ = coloured seed and green plant

$r^r = sP$ = colourless seed and coloured (red) plant

$rS = sp$ = colourless seed and green plant

The new designations were based on the assumption that *S* and *P* are two closely linked genes rather than forming a series of multiple alleles. But since alleles later called pseudoalleles, can undergo crossing over between them, the concept of multiple alleles may apply in this case.

METHOD- STUDENT ACTIVITY

Inspecting the Corn Kernels

1. Inspect the first generation (F1) offspring. This has been produced via a cross of 1purple kernel and 1 yellow kernel parent.

2. Record your observations.
3. Inspect the second generation (F2) of corn and count the number of each different phenotype exhibited in the colour of the kernel. Any purple at all must be recorded as purple.
4. Add your count of F2 corn kernels data.
5. Keep a record of the class data for analysis

Making Punnett Squares

1. List the genotype for the parent and F1 generations, using R to represent purple colouring and r to represent yellow.
2. Using the genotype for the parent and F1 generations, perform a cross between a homozygous purple parent and yellow parent. Use a Punnett square to illustrate your answers.
3. Perform a cross between two F1 offspring using a Punnett square.
4. Calculate the ratio of phenotypes of the F2 generation and compare the calculated class phenotype ratio and the F2 generation ratio

OBSERVATION AND RESULTS

Below is a summary of the expected results of the above procedure. This is to be used as a guide only, as individual results will vary.

	Phenotype (Colour)	Phenotype (Colour)
F_1	All purple	
F_2	Purple	Yellow

Table : Corn Kernel Phenotype

F_2	Phenotype (Colour)	Phenotype (Colour)
Group	180 Purple	60 Yellow
Class totals	1442 Purple	478 Yellow
Class Percent	(1442/1920) x 100 =75%	(478/1920) x 100 =25%

Table : Count of F2 Corn Kernels

	R	R
r	Rr	Rr
r	Rr	Rr

Figure : Cross between a Homozygous Purple Parent and Yellow Parent

	R	r
R	RR purple	Rr Purple
r	Rr Purple	Rr yellow

Figure : Cross between Two F1 Offspring

Ratio of Purple to yellow	75/25=3:1or 144/478=3:1

Figure : Ratio of Phenotypes of the F2 Generation

INVESTIGATION

- Ask your students to consider if there is any evidence to support the idea that one of the phenotypes is dominant.
- Based on the F2 generation data, ask students which phenotype exhibits recessive traits. Students should provide evidence for their answers.
- Challenge your students to identify which of Mendel's laws is being investigated in this practical and provide evidence to support their answers.
- Encourage your students to explain the differences or similarities between Mendel's laws.

Skin color in rodents

In rabbits, four kinds of skin colour are known. Rabbits are accordingly classified as **coloured (agouti), chinchilla, himalayan albino,** and **albino.** If homozygous coloured *(CC)* is crossed with albino (c^ac^a), F_1heterozygous (Cc^a) individual is coloured showing that coloured is dominant over albino. F_1 heterozygous coloured (Cc^a) gives a 3 : 1 ratio in F_2.

When coloured *(CC)* are crossed with himalayan albino (c^hc^h), F_1 individuals are coloured (Cc^h). These in turn give 3 : 1 ratio between coloured and himalayan albino. Similar relationship is observed when himalayan albino (having pink eyes and white coat colour except at the tips like nose, tail and feet, etc. which are black) is crossed with albino.

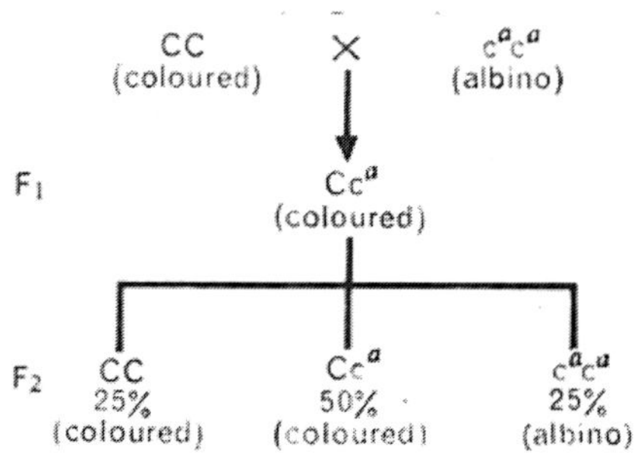

Fig.. Inheritance of skin in rabbits from a cross (coloured x albino).

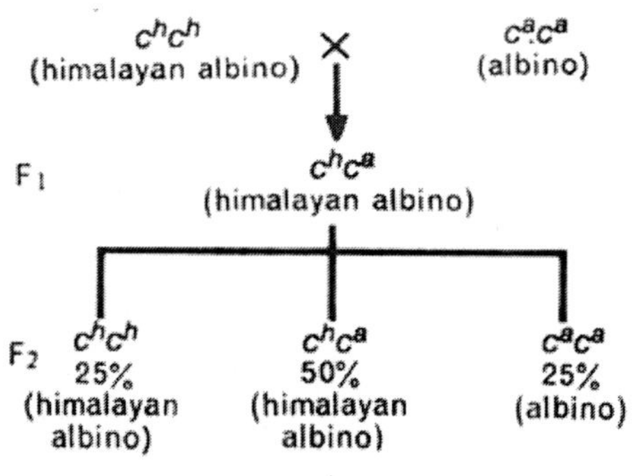

Fig. Inheritance of skin colour in rabbits from a cross (Himalayan albino x albino).

Still another type is known in rabbits and is called **chinchilla.** This colour is lighter than the wild (agouti). Coloured character is completely dominant over chinchilla. However F_1hybrids between chinchilla and.himalayan albino $(c^{ch}c^h)$ or between chinchilla and albino

$(c^{ch}c^{a})$ show light gray skin colour. A summary of different genotypes involved in this series, along with their phenotypes.

Table. Various genotypes and their phenotypes, due to multiple allelic series for skin colour in rabbit

Genotypes	Phenotypes
CC, Cc^{ch}, Cc^{a}	coloured (wild)
$c^{ch}c^{ch}$	chinchilla
$c^{ch}c^{h}$, $c^{ch}c^{a}$	light grey
$c^{h}c^{h}$, $c^{h}c^{a}$	himalayan albino
$c^{a}c^{a}$	albino

In mice, however, two different multiple allelic series are known for skin colour, one is albino series and other a black series. In albino series, it was observed that wild type (grey) was dominant over all others, while heterozygous between any two of the three recessive alleles. $(a, a^{m}$ and $a^{e})$ gave intermediate phenotype. In black series on the other hand, black is recessive to all, while yellow is dominant to all. Light belly in G^{L} and b^{L} is dominant over darker belly. Black body of b^{L}, however, is recessive to grey body of normal individuals. Consequently, $+/G^{L}$ and $+lb^{L}$ both would have grey body and light belly.

Table. Various homozyfous genotypes and phenotypes, due to two different multiple allelic series for skin colour in mice.

Albino series		Black series	
Genotype	**Phenotype**	**Genotype**	**Phenotype**
$+/+$	grey	$+/+$	normal grey
a/a	albino	b/b	black
a^{m}/a^{m}	medium light	Y/Y	yellow-dies (lethal)
a^{e}/a^{e}	extreme light	G^{L}/G^{L}	grey with light belly
		b^{L}/b^{L}	black with light belly

Table. Alleles for albino series in mice, in various heterozygous comhinations and their corresponding phenotypes.

Genotype	Phenotype
$+/a$ $+/a^m$ $+/a^e$	grey
a/a^m	interm. between a/a and a^m/a^m
a/a^e	interm. between a/a and a^e/a^e
a^m/a^e	interm. between a^m/a^m and a^e/a^e

Table. Alleles for black series in mice, in various heterozygous comhinations and their corresponding phenotypes.

Genotype	Phenotype
$+/b$	grey
$+/Y$ Y/b	yellow
$+/G^L$ $+/b^L$ G^L/b^L G^L/b	grey body with light belly
b^L/b	black body with light belly

Eye color in drosophila

Drosophila melanogaster is an extremely useful model organism for both classical genetics and molecular biology. It is also one of the earliest organisms to have its DNA sequence determined and extensively annotated. However, many of the genes identified and used in classical genetics have not been matched with their corresponding genes in the molecular sequence. This study identifies four genes coding for eye color mutations.

Certain eye color genes regulate vesicular transport in cells. Enzymes and other substances needed for pigment synthesis are transported to the pigment granule, a lysosome-related organelle. *D. melanogaster* has a red-brown eye color caused by the presence of two classes of pigments, pteridines (red) and ommochromes (brown). There are independent pathways for the synthesis of each (Reaume *et al.* 1991; Summers *et al.* 1982). Some eye color mutations are caused by the lack of individual enzymes in these pathways. Examples include bright red *vermillion* (tryptophan 2,3-dioxygenase), which is incapable of synthesis of any ommochromes, and dark brown *sepia* (GSTO4), which lacks an enzyme used to synthesize a subset of pteridines (Baillie and Chovnick 1971; Kim *et al.* 2006; Searles and Voelker 1986; Searles *et al.* 1990; Walters *et al.* 2009; Wiederrecht and Brown 1984). The first mutant described in *Drosophila, white,* lacks both types of pigments in the eye (Morgan 1910). The *white* gene codes for a subunit of an ABC transporter located in the membrane of the pigment granule, and is required for transport of substrates into pigment granules (Goldberg *et al.* 1982; Mackenzie *et al.* 1999; Summers *et al.* 1982). Lloyd *et al.* (1998) developed a hypothesis that single mutations causing perturbations in the amount of both types of pigments in the eyes could result from defects in genes associated with intracellular transport. They named these genes "the granule group." Their concept has been supported by both identification of existing eye color genes for which the genomic coding sequences were originally unknown, and by studies in which induced mutations or reduced expression of orthologs of known transporter genes resulted in defective eye color phenotypes.

Normal eye colour in *Drosophila* is red. White eyed *Drosophila* was one of the first mutations known in *Drosophila.* Red eye and white eye showed simple dominant recessive relationship.

Table. Multiple alleles for eye colour in Drosophila and various phenotypes due to homozygous and heterozygous conditions.

Allele	Phenotype in homozygous condition	Heterozygotes	
		Genotypes	Phenotypes
+	red	$+/w$	red eyes
w	white	$+/w^{ch}$	
w^{ch}	cherry	$+/w^{b}$	
w^{b}	blood	$+/w^{e}$	
w^{e}	eosin	$+/w^{apr}$	
w^{apr}	apricot	$+/w^{iv}$	
w^{iv}	ivory	$+/w^{cr}$	
w^{cr}	cream	other heterozygotes	intermediate

Subsequently, different shades between red and white were recovered. About a dozen different allelomorphs are known now at this locus. Red eye was dominant over all others. When any two recessive alleles were brought together, intermediate type was obtained.

Self sterility in nicotiana blood group in Humans - ABO typing - Rh factor

Blood Groups in Man:

Several genes in man produce multiple allelic series which affect an interesting and important physiological characteristic of the human red blood cells. The red blood cells have special antigens properties by which they respond to certain specific components (antibodies) of the blood serum.

The antigen-antibody relationship is one of the great specificity like that between lock and key. Each antigen and its associated antibody has a peculiear chemical configuration. Landsteiner discovered in 1900 that when the red cells of one person are placed in the blood serum of another person, the cells become clumped or agglutinated.

If blood transfusions were made between persons of two such incompatible blood groups, the transfused cells were likely to clump and shut out the capillaries in the recipient, some times resulting in death.

However, such reactions occurred only when the cells of certain individuals were placed in serum from certain other persons. It was found that all persons could be classified in to four groups with regard to the antigen property of the blood cells.

Large number of persons have been classified in to these four groups by means of the agglutination test and the distribution of blood groups in the offspring of parents of known blood groups has been studied. The evidence shows that these blood properties are determined by a series of three allelic genes I^A, I^B and i, as follows:

Blood groups	Genotype
AB	I^AI^B
B	I^BI^B or I^Bi
A	I^AI^A or I^Ai
O	ii

I^A is a gene for the production of the anti-gin A. I^B for antigen B, and i for neither antigen. The existence of these alleles in man and the case with which the blood groups can be identified have obvious practical applications in blood transfusion, cases of disputed percentage and description of human populations.

The alleles of these genes which affect a variety of biochemical properties of the blood, act in such a way that in the heterozygous compound I^AI^B, each allele exhibits its own characteristics and specific effect. The cells of the heterozygote contain both antigens A and B. On the other hand, I^A and I^B both show complete dominance over i, which lacks both antigens.

PARENTS		CHILDREN	
Phenotypes	Genotypes	Phenotypes	Genotypes
O X O	ii X ii	O	ii
O X A	ii X I^AI^A or I^Ai	O, A	ii, I^Ai
O X B	ii X I^BI^B or I^Bi	O, B	ii, I^Bi
O X AB	ii X I^AI^B	A, B	I^Ai, I^Bi
A X A	I^AI^A or I^Ai X I^AI^A or I^Ai	A, O	I^AI^A, ii
A X B	I^AI^A or I^Ai X I^BI^B or I^Bi	A, AB, O, B	I^Ai, I^AI^B, ii, I^Bi
A X AB	I^AI^A or I^Ai X I^AI^B	A, B, AB	I^AI^A, I^Bi, I^AI^B
B X B	I^BI^B or I^Bi X I^BI^B or I^Bi	B, O	I^BI^B, ii
B X AB	I^BI^B or I^Bi X I^AI^B	A, B, AB	I^AI^A, I^BI^B, I^AI^B
AB X AB	I^AI^B X I^AI^B	A, B, AB	I^AI^A, I^BI^B, I^AI^B

Table showing possible blood types of children from parents of various blood groups.

The 'Rhesus' Blood Group in Man:

A very interesting series of alleles affecting the antigens of human blood has been discovered through the work of Landsteiner, Wiener, Race, Levine, Sanger, Mourant & several others.

The original discovery was that the red cells are agglutinated by a serum prepared by immunizing rabbits against the blood of Rhesus monkey. The antigen responsible for this reaction was consequently called as Rhesus factor and the gene that causes this property was denoted as R-r or Rh-rh.

Interest in this factor was stimulated by Levine's study of a characteristic form of anaemia, known as Erythroblastosis foetalis, which occurs occasionally in new born infants.

It was found that the infants suffering from this anaemia are usually Rh-positive and so are their fathers; but their mothers are Rh-negative. The origin of the disease was explained as follows: The Rh^+ foetus developing in the uterus of an Rh^- mother causes the formation of mother's blood stream of anti Rh antibodies.

These antibodies, especially as a result of a succession of several Rh^+ pregnancies, gain sufficient strength in the mother's blood so that they may attack the red blood cells of the foetus. The reaction between these antibodies of the mother and the red cells of her unborn child provokes haemolysis and anaemia; this may be serious enough to cause the death of the newborn infant or abortion of the foetus.

The blood stream of a mother who has had an erythroblastotic infant is a much more potent and convenient reagent than sera of rabbits, immunized by blood of rhesus monkey's for testing the blood of other persons to distinguish Rh^+ from Rh^- individuals using such sera from woman who had erythroblastotic infants, it was discovered that: there exist not one but several kinds of Rh^+ and Rh- persons. There are several different Rh antigens which are detected by specific antisera.

Thus, an Rh^- woman immunized during pregnancy by the Rh^+ children may have in her blood serum antibodies, that agglutinate not only Rh+ red cells but also cells from a few persons known to be Rh^-.

By selective absorption two kinds of antibodies may be separated from such a serum, one known as anti-D which agglutinates (= coagulates) only Rh^+ cells, the other known as anti-C which agglutinates particular rare types of Rh^-. Another specific antibody, known as anti-c agglutinates all cells that lack C.

With these three antisera, six types of blood can be recognized. Studies of parent and children show that persons of type Cc are heterozygous for an allele C determining C anti-gena. CC persons are homozygous for C and cc are homozygous for c. There is obviously no dominance, each allele producing its own antigen in the heterozygote as in the AB blood type.

No anti serum is available for detecting d, the alternative to D. D^+ persons may be heterozygous or homozygous. However, the genotypes of such persons may be diagnosis from their progeny; for example D^+ person who has a d^- child is thereby shown to be Dd.

Two other specific antibodies, anti-E and anti-c have been found. These detect the antigens E and e determined by a pair of alleles E and e. The three elementary types of antigens C-c, D and E-e, occur in fixed combinations that are always inherited together as alleles of a single gene. Wiener and Fisher showed the existence of a series of eight different alternative arrangements of these three types of Rh antigens and expressed them by means of following symbols.

The Rh System of Alleles:

Fisher's symbols	Wiener's symbols	
CDE	R^z	Rh-Positive
CDe	R^1	
cDE	R^2	
cDe	R^0	
CdE	r^Y	Rh-Negative
Cde	r'	
cdE	r''	
cde	r	

Thus, allelism is determined by cross-breeding experiments. If one gene behaves as dominant to another the conclusion is that they are alleles and that they occupy identical loci in homologous chromosomes when two genes behave as dominant to other gene. They should occupy identical loci in the chromosome. When more than a pair of alleles occur in respect of any character in inheritance the phenomenon is known as multiple allelism.

There is not much difference between the two theories of Wiener and Fisher. Wiener opinion is that there are multiple variations of one gene whereas according to the view of Fisher three different genes lying very close together are responsible for differences.

Pleiotropism:

The opposite of polygene effect is known as pleiotropism i.e., a single gene influence or govern many characters. For example, gene for vestigial wing influence the nature of halters (modified balancers of Drosophila). The halters are not normal but reduced in flies with vestigial wings. The vestigial gene also affects position of dorsal bristles which instead of being horizontal turn out to be vertical.

This gene also affects the shape of spermatheca i.e., the shape of spermatheca is changed ; the number of egg strings in the ovaries is decreased compared to normal when the vestigial larvae are well fed but relatively increased when they are poorly fed ; length of life and fruitfulness or fertility are lowered, and there are still other differences.

ABO blood groups in humans

On the basis of presence or absence of certain antigens, ABO blood groups have been established in man by **K. Landsteiner. It** was found that there can be two antigens A or B and as a result four blood groups. These blood groups are called ABO blood groups or Landsteiner blood groups after the name of the discoverer.

With these antigens A and B, there are certain naturally occurring antibodies in the serum of the blood. General principle of antibody and antigen relationships is, that antibodies in a particular individual will be found only for those antigens which are absent in blood of this individual. The status of antigens and antibodies in different blood groups.

Table : Different ABO blood groups and the antigens and antibodies associated with them

Blood groups	Antigen	Antibody for antigen
A	A	B
B	B	A
AB	A and B	None
O	None	A and B

Antibodies in blood group A will be able to agglutinize red blood corpuscles of the blood group B and *vice versa.* AB blood group will not agglutinize any other group, since no antibodies are present. Similarly,

O blood group should be able to agglutinize all other three groups except its own. On this relationship is based the test whether a particular blood group can be donated for another specific blood group.

If donor's blood has antibodies against that of recipient, it does not matter, since they get diluted. However, there should be no antibodies in recipient's blood against donor's blood, since this will cause agglutinization of transfused blood. It would also be obvious, that AB is **universal recipient** and no test of donor's blood in such case is necessary. Similarly, O group is **universal donor** and no test for the recipient's blood is necessary, if blood of O group is available for transfusion.

Table: Compatible donor and recipient ABO blood Groups

Blood group of donor	Blood group of recipient
A	A and AB
B	B and AB
AB	AB (universal recipient)
O (universal donor)	O, A, B, AB

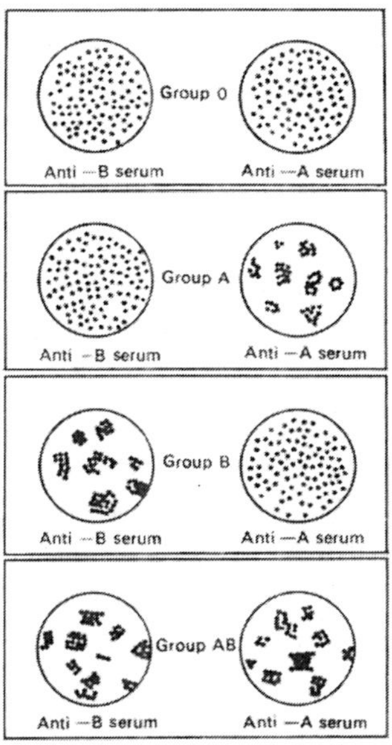

Fig. Effects of anti-A (containing antibody A) and anti-B (containing antibody B) sera on corpuscles of four blood groups (O, A, B and AB). A clumped pattern within a circle shows agglutinization of red blood corpuscles.

In ABO blood groups discussed above, **antigens** (also called **agglutinogens)** are present in the red blood cells and **antibodies** (also called **agglutinins)** in the serum. Consequently, transfusions can be made after examining the compatibilities of blood groups, so that the transfused blood may not get agglutinated. As shown, the blood group A contains anti-B serum (antibodies for B antigen) and blood group B contains anti-A serum. The effects of anti-A serum and anti-B serum on agglutinization of different blood groups.

The genetics of these blood groups indicates that three alleles are present (i) *1° or i* or + (ii) I^A *or A* and (iii) I^B or *B*. l^A and I^B are mutant alleles and are dominant over the wild allele *1°* or *i* or +. This is based on the concept that wild type is one which is more frequent. O and AB blood groups are more frequent and if O is assumed as wild, it yields a convenient system. The genotypes of different blood groups.

Table: Different ABO blood groups and their genotypes

Blood group	Possible genotypes
O	I^OI^O or *ii* or ++
A	I^AI^A or *AA*; I^AI^O or I^A *i* or *A*/+
B	I^BI^B or *BB*; I^BI^O or I^B *i* or *B*/+
AB	I^AI^B or *A*/*B*

MN series of blood groups in humans

Besides series of ABO blood groups, there are other series like MN blood groups. In case of MN series, human blood serum does not contain any specific antibodies, which will cause agglutinization of any blood group in this series. But if the human blood is injected in rabbit, it causes production of specific antibodies responsible for agglutinization of blood type injected. For instance, if blood of M group is injected in rabbit, it will cause production of antibodies in serum (anti-M serum) causing agglutinization of M as well as MN but not of N. Reverse will be the case, when blood of N type is injected in rabbit, so that anti-N serum produced in rabbit causes agglutinization of N and MN, but not of M.

Landsteiner and Levine (1927) divided human populations into three groups on the basis of reaction with anti-M serum and anti-N serum produced in rabbits. In honour of Landsteiner, the gene for this MN series was named as *L* and alleles as L^M and L^N, also now called

M and N. These alleles are co-dominant, so that the heterozygote $L^M L^N$ gives rise to the blood group MN.

Table : Different MN Blood groups, their genotypes, antigens and reactions with anti-M and anti-N sera

Blood group	Genotypes	Cellular antigen	Serum anti-body	Reaction with serum*	
				Anti-M serum	Anti-N serum
M	$L^M L^M$ or MM	M	None	+	–
MN	$L^M L^N$ or MN	M, N	None	+	+
N	$L^N L^N$ or NN	N	None	–	+

* + means agglutinization; serum is rabbit immunized.

New varieties of MN blood types were discovered by **Walsh** and **Montgomery** in 1947. A new blood-type system S_3 was shown to be closely tied to the inheritance of MN gene system. Thus four alleles, namely L^{MS}, L^{Ms}, L^{NS}, L^{Ns} (or M^S M^s, N^S, N^s) could be identified and nine blood groups were possible.

Table : Nine possible MNSs blood groups and reactions with four different anti sera

Blood group	Genotypes	Reaction with immunized serum			
		anti-M	anti-N	anti-S	anti-s
MS	$L^{MS}L^{MS}$ or M^SM^S	+	–	+	–
MSs	$L^{MS}L^{Ms}$ or M^SM^s	+	–	+	+
Ms	$L^{Ms}L^{Ms}$ or M^sM^s	+	–	–	+
MNS	$L^{MS}L^{NS}$ or M^SN^S	+	+	+	–
MNSs	$L^{MS}L^{Ns}$ or M^SN^s or $L^{Ms}L^{NS}$ or M^sN^S	+	+	+	+
MNs	$L^{Ms}L^{Ns}$ or M^sN^s	+	+	–	+
NS	$L^{NS}L^{NS}$ or N^SN^S	–	+	+	–
NSs	$L^{NS}L^{Ns}$ or N^SN^s	–	+	+	+
Ns	$L^{Ns}L^{Ns}$ or N^sN^s	–	+	–	+

Blood groups and disputed parentage

Blood groups sometimes help to decide cases of disputed parentage in criminal courts, because a particular pair of blood groups in parents may give some and not all blood groups in progeny. Therefore, if a child has a blood group, which is not likely to result from the blood groups of a married couple claiming parentage, then it is proved that the child has a doubtful parentage or in other words is illegitimate. Various possible pairs of blood groups of the parents, their children and the impossible blood groups are given in In recent years, however, technique of DNA fingerprinting has become available to decide cases of disputed parentage.

Table : Different mating combinations of ABO blood groups, their resulting progenies and the improbable blood groups

Parents with blood groups and genotypes	Progeny with blood groups and genotypes	Improbable blood groups
1. O (I^OI^O) × O (I^OI^O)	O (I^OI^O)	A, B, AB
2. O (I^OI^O) × A (I^AI^A, I^AI^O)	O (I^OI^O), A (I^AI^O)	B, AB
3. O (I^OI^O) × B (I^BI^B, I^BI^O)	O (I^OI^O), B (I^BI^O)	A, AB
4. O (I^OI^O) × AB (I^AI^B)	A (I^AI^O), B (I^BI^O)	O, AB
5. A (I^AI^A, I^AI^O) × A (I^AI^A, I^AI^O)	A (I^AI^A, I^AI^O), O (I^OI^O)	B, AB
6. A (I^AI^A, I^AI^O) × B (I^BI^B, I^BI^O)	A (I^AI^O), B (I^BI^O), AB (I^AI^B), O (I^OI^O)	none
7. A (I^AI^A, I^AI^O) × AB (I^AI^B)	A (I^AI^A, I^AI^O), B (I^BI^O), AB (I^AI^B)	O
8. B (I^BI^B, I^BI^O) × AB (I^AI^B)	A (I^AI^O), B (I^BI^B, I^BI^O), AB (I^AI^B)	O
9. AB (I^AI^B) × AB (I^AI^B)	A (I^AI^A), B (I^BI^B), AB (I^AI^B)	O

Table : Different mating combinations of MN blood groups, resulting progenies and improbable blood groups

Parents with blood groups and genotypes	Progeny with blood groups and genotypes	Improbable blood groups
M (L^ML^M) × M (L^ML^M)	M (L^ML^M)	N, MN
M (L^ML^M) × N (L^NL^N)	MN (L^ML^N)	M, N
M (L^ML^M) × MN (L^ML^N)	1 M (L^ML^M) : 1 MN (L^ML^N)	N
N (L^NL^N) × N (L^NL^N)	N (L^NL^N)	M, MN
N (L^NL^N) × MN (L^ML^N)	1 N (L^NL^N) : 1 MN (L^ML^N)	M
MN (L^ML^N) × MN (L^ML^N)	1 M (L^ML^M) : 2 MN (L^ML^N) : 1 N (L^NL^N)	None

Rh blood groups and erythroblastosis fetalis

Rh factor was discovered by **Landsteiner** and **Wiener** in 1940 from rabbits immunized with the blood of **rhesus monkeys** *(Macacus rhesus).* The resulting antibodies in rabbit serum agglutinated blood of not only rhesus monkeys, but also of certain percentage of human beings and these human beings were called *Rh*+ (*Rh* positive). The *Rh* antigen can produce antibodies in human serum also, which may be possible through wrong transfusion of blood. Therefore, to avoid agglutinization, cross-compatibility of *Rh* factor as well as ABO blood groups is necessary before transfusion of blood is made. *Rh* negative individuals should always be given *Rh*-negative blood to avoid subsequent reaction due to antibody formation.

Chapter 5

Chromosomal Theory of Inheritance

Introduction to Chromosomes

The speculation that chromosomes might be the key to understanding heredity led several scientists to examine Mendel's publications and re-evaluate his model in terms of the behavior of chromosomes during mitosis and meiosis. In 1902, Theodor Boveri observed that proper embryonic development of sea urchins does not occur unless chromosomes are present. That same year, Walter Sutton observed the separation of chromosomes into daughter cells during meiosis. Together, these observations led to the development of the Chromosomal Theory of Inheritance, which identified chromosomes as the genetic material responsible for Mendelian inheritance.

The Chromosomal Theory of Inheritance was consistent with Mendel's laws and was supported by the following observations:

- During meiosis, homologous chromosome pairs migrate as discrete structures that are independent of other chromosome pairs.
- The sorting of chromosomes from each homologous pair into pre-gametes appears to be random.
- Each parent synthesizes gametes that contain only half of their chromosomal complement.
- Even though male and female gametes (sperm and egg) differ in size and morphology, they have the same number of chromosomes, suggesting equal genetic contributions from each parent.
- The gametic chromosomes combine during fertilization to produce offspring with the same chromosome number as their parents.

Despite compelling correlations between the behavior of

chromosomes during meiosis and Mendel's abstract laws, the Chromosomal Theory of Inheritance was proposed long before there was any direct evidence that traits were carried on chromosomes. Critics pointed out that individuals had far more independently segregating traits than they had chromosomes. It was only after several years of carrying out crosses with the fruit fly, *Drosophila melanogaster*, that Thomas Hunt Morgan provided experimental evidence to support the Chromosomal Theory of Inheritance.

In 1910, Thomas Hunt Morgan started his work with *Drosophila melanogaster*, a fruit fly. He chose fruit flies because they can be cultured easily, are present in large numbers, have a short generation time, and have only four pair of chromosomes that can be easily identified under the microscope. They have three pair of autosomes and a pair of sex chromosomes. At that time, he already knew that X and Y have to do with gender. He used normal flies with red eyes and mutated flies with white eyes and cross bred them. In flies, the wild type eye color is red (XW) and is dominant to white eye color (Xw). He was able to conclude that the gene for eye color was on the X chromosome. This trait was thus determined to be X-linked and was the first X-linked trait to be identified. Males are said to be hemizygous, in that they have only one allele for any X-linked characteristic.

Figure : Eye Color in Fruit Flies: In Drosophila, the gene for eye color is located on the X chromosome. Red eye color is wild type and is dominant to white eye color.

Association of maternal & paternal chromosomes of meiosis

Meiosis

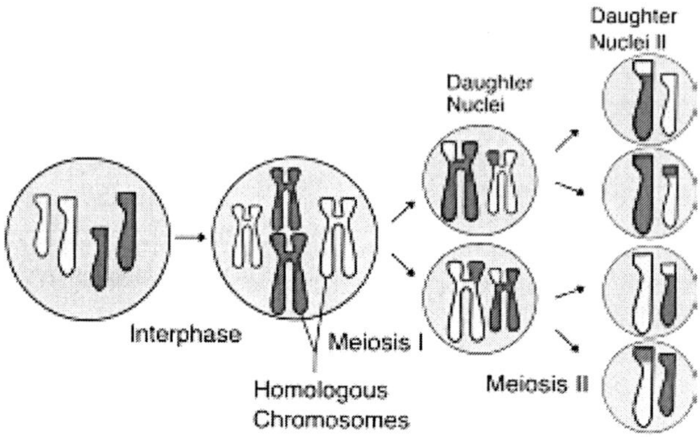

In meiosis, the chromosomes duplicate (during interphase) and homologous chromosomes exchange genetic information (chromosomal crossover) during the first division, called meiosis I. The daughter cells divide again in meiosis II, splitting up sister chromatids to form haploid gametes. Two gametes fuse during fertilization, forming a diploid cell (zygote) with a complete set of paired chromosomes.

Meiosis 'lessening', since it is a reductional division) is a special type of cell division of germ cells in sexually-reproducing organisms that produces the gametes, such as sperm or egg cells. It involves two rounds of division that ultimately result in four cells with only one copy of each chromosome (haploid). Additionally, prior to the division, genetic material from the paternal and maternal copies of each chromosome is crossed over, creating new combinations of code on each chromosome. Later on, during fertilisation, the haploid cells produced by meiosis from a male and female will fuse to create a cell with two copies of each chromosome again, the zygote.

Errors in meiosis resulting in aneuploidy (an abnormal number of chromosomes) are the leading known cause of miscarriage and the most frequent genetic cause of developmental disabilities.

In meiosis, DNA replication is followed by two rounds of cell division to produce four daughter cells, each with half the number of chromosomes as the original parent cell. The two meiotic divisions are known as meiosis I and meiosis II. Before meiosis begins, during S phase of the cell cycle, the DNA of each chromosome is replicated so that it consists of two identical sister chromatids, which remain held together through sister chromatid cohesion. This S-phase can be referred to as "premeiotic S-phase" or "meiotic S-phase". Immediately following DNA replication, meiotic cells enter a prolonged G_2-like stage known as meiotic prophase. During this time, homologous chromosomes pair with each other and undergo genetic recombination, a programmed process in which DNA may be cut and then repaired, which allows them to exchange some of their genetic information. A subset of recombination events results in crossovers, which create physical links known as chiasmata (singular: chiasma, for the Greek letter Chi (×)) between the homologous chromosomes. In most organisms, these links can help direct each pair of homologous chromosomes to segregate away from each other during meiosis I, resulting in two haploid cells that have half the number of chromosomes as the parent cell.

During meiosis II, the cohesion between sister chromatids is released and they segregate from one another, as during mitosis. In some cases, all four of the meiotic products form gametes such as sperm, spores or pollen. In female animals, three of the four meiotic products are typically eliminated by extrusion into polar bodies, and only one cell develops to produce an ovum. Because the number of chromosomes is halved during meiosis, gametes can fuse (i.e. fertilization) to form a diploid zygote that contains two copies of each chromosome, one from each parent. Thus, alternating cycles of meiosis and fertilization enable sexual reproduction, with successive generations maintaining the same number of chromosomes. For example, diploid human cells contain 23 pairs of chromosomes including 1 pair of sex chromosomes (46 total), half of maternal origin and half of paternal origin. Meiosis produces haploid gametes (ova or sperm) that contain one set of 23 chromosomes. When two gametes (an egg and a sperm) fuse, the resulting zygote is once again diploid, with the mother and father each contributing 23 chromosomes. This same pattern, but not the same number of chromosomes, occurs in all organisms that utilize meiosis.

Meiosis occurs in all sexually-reproducing single-celled and multicellular organisms (which are all eukaryotes), including animals, plants and fungi. It is an essential process for oogenesis and spermatogenesis.

Overview

Although the process of meiosis is related to the more general cell division process of mitosis, it differs in two important respects:

Recombination	meiosis	shuffles the genes between the two chromosomes in each pair (one received from each parent), producing recombinant chromosomes with unique genetic combinations in every gamete
	mitosis	occurs only if needed to repair DNA damage; usually occurs between identical sister chromatids and does not result in genetic changes
Chromosome number (ploidy)	meiosis	produces four genetically unique cells, each with half the number of chromosomes as in the parent
	mitosis	produces two genetically identical cells, each with the same number of chromosomes as in the parent

Meiosis begins with a diploid cell, which contains two copies of each chromosome, termed homologs. First, the cell undergoes DNA replication, so each homolog now consists of two identical sister chromatids. Then each set of homologs pair with each other and exchange genetic information by homologous recombination often leading to physical connections (crossovers) between the homologs. In the first meiotic division, the homologs are segregated to separate daughter cells by the spindle apparatus. The cells then proceed to a second division without an intervening round of DNA replication. The sister chromatids are segregated to separate daughter cells to produce a total of four haploid cells. Female animals employ a slight variation on this pattern and produce one large ovum and two small polar bodies. Because of recombination, an individual chromatid can consist of a new combination of maternal and paternal genetic information, resulting in offspring that are genetically distinct from either parent. Furthermore, an individual gamete can include an assortment of maternal, paternal, and recombinant chromatids. This genetic diversity resulting from sexual reproduction contributes to the variation in traits upon which natural selection can act.

Meiosis uses many of the same mechanisms as mitosis, the type of cell division used by eukaryotes to divide one cell into two identical daughter cells. In some plants, fungi, and protists meiosis results in the formation of spores: haploid cells that can divide vegetatively without undergoing fertilization. Some eukaryotes, like bdelloid rotifers, do not have the ability to carry out meiosis and have acquired the ability to reproduce by parthenogenesis.

Meiosis does not occur in archaea or bacteria, which generally reproduce asexually via binary fission. However, a "sexual" process known as horizontal gene transfer involves the transfer of DNA from one bacterium or archaeon to another and recombination of these DNA molecules of different parental origin.

History

Meiosis was discovered and described for the first time in sea urchin eggs in 1876 by the German biologist Oscar Hertwig. It was described again in 1883, at the level of chromosomes, by the Belgian zoologist Edouard Van Beneden, in *Ascaris* roundworm eggs. The significance of meiosis for reproduction and inheritance, however, was described only in 1890 by German biologist August Weismann, who noted that two cell divisions were necessary to transform one diploid cell into four haploid cells if the number of chromosomes had to be maintained. In 1911, the American geneticist Thomas Hunt Morgan detected crossovers in meiosis in the fruit fly *Drosophila melanogaster*, which helped to establish that genetic traits are transmitted on chromosomes.

The term "meiosis" is derived from the Greek word ìåßùóćò, meaning 'lessening'. It was introduced to biology by J.B. Farmer and J.E.S. Moore in 1905, using the idiosyncratic rendering "maiosis":

We propose to apply the terms Maiosis or Maiotic phase to cover the whole series of nuclear changes included in the two divisions that were designated as Heterotype and Homotype by Flemming.

The spelling was changed to "meiosis" by Koernicke (1905) and by Pantel and De Sinety (1906) to follow the usual conventions for transliterating Greek.

Phases

Meiosis is divided into meiosis I and meiosis II which are further divided into Karyokinesis I, Cytokinesis I, Karyokinesis II, and Cytokinesis II, respectively. The preparatory steps that lead up to meiosis are identical in pattern and name to interphase of the mitotic cell cycle. Interphase is divided into three phases:

- Growth 1 (G_1) phase: In this very active phase, the cell synthesizes its vast array of proteins, including the enzymes and structural proteins it will need for growth. In G_1, each of the chromosomes consists of a single linear molecule of DNA.
- Synthesis (S) phase: The genetic material is replicated; each of the cell's chromosomes duplicates to become two identical sister chromatids attached at a centromere. This replication does not change the ploidy of the cell since the centromere number remains the same. The identical sister chromatids have not yet

condensed into the densely packaged chromosomes visible with the light microscope. This will take place during prophase I in meiosis.

- Growth 2 (G_2) phase: G_2 phase as seen before mitosis is not present in meiosis. Meiotic prophase corresponds most closely to the G_2 phase of the mitotic cell cycle.

Interphase is followed by meiosis I and then meiosis II. Meiosis I separates replicated homologous chromosomes, each still made up of two sister chromatids, into two daughter cells, thus reducing the chromosome number by half. During meiosis II, sister chromatids decouple and the resultant daughter chromosomes are segregated into four daughter cells. For diploid organisms, the daughter cells resulting from meiosis are haploid and contain only one copy of each chromosome. In some species, cells enter a resting phase known as interkinesis between meiosis I and meiosis II.

Meiosis I and II are each divided into prophase, metaphase, anaphase, and telophase stages, similar in purpose to their analogous subphases in the mitotic cell cycle. Therefore, meiosis includes the stages of meiosis I (prophase I, metaphase I, anaphase I, telophase I) and meiosis II (prophase II, metaphase II, anaphase II, telophase II).

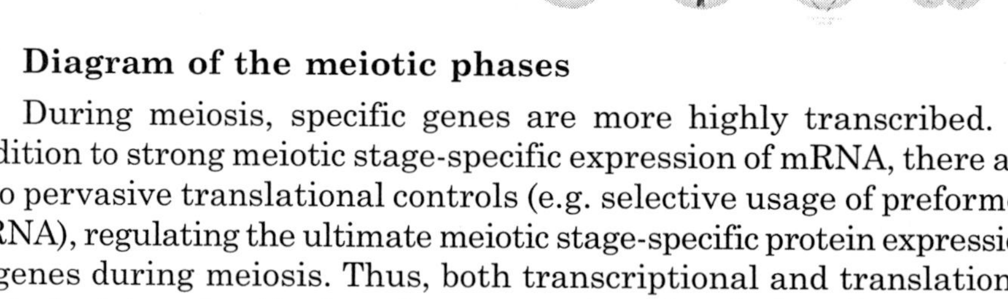

Diagram of the meiotic phases

During meiosis, specific genes are more highly transcribed. In addition to strong meiotic stage-specific expression of mRNA, there are also pervasive translational controls (e.g. selective usage of preformed mRNA), regulating the ultimate meiotic stage-specific protein expression of genes during meiosis. Thus, both transcriptional and translational controls determine the broad restructuring of meiotic cells needed to carry out meiosis.

Meiosis I

Meiosis I segregates homologous chromosomes, which are joined as tetrads (2n, 4c), producing two haploid cells (n chromosomes, 23 in humans) which each contain chromatid pairs (1n, 2c). Because the ploidy is reduced from diploid to haploid, meiosis I is referred to as a *reductional division*. Meiosis II is an *equational division* analogous to mitosis, in which the sister chromatids are segregated, creating four haploid daughter cells (1n, 1c).

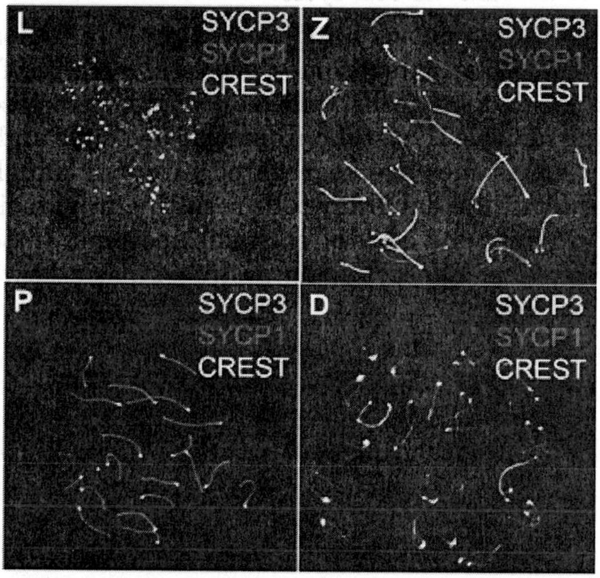

Meiosis Prophase I in mice. In Leptotene (L) the axial elements (stained by SYCP3) begin to form. In Zygotene (Z) the transverse elements (SYCP1) and central elements of the synaptonemal complex are partially installed (appearing as yellow as they overlap with SYCP3). In Pachytene (P) it's fully installed except on the sex chromosomes. In Diplotene (D) it disassembles revealing chiasmata. CREST marks the centromeres.

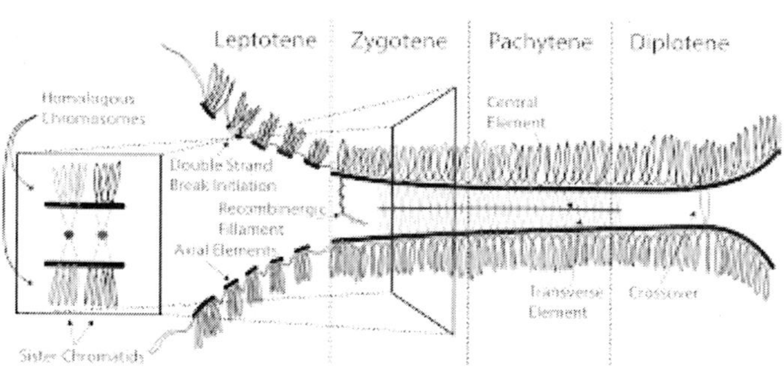

Schematic of the synaptonemal complex at different stages of prophase I and the chromosomes arranged as a linear array of loops.

Prophase I

Prophase I is by far the longest phase of meiosis (lasting 13 out of 14 days in mice). During prophase I, homologous maternal and paternal chromosomes pair, synapse, and exchange genetic information (by homologous recombination), forming at least one crossover per chromosome. These crossovers become visible as chiasmata (plural; singular chiasma). This process facilitates stable pairing between homologous chromosomes and hence enables accurate segregation of the chromosomes at the first meiotic division. The paired and replicated chromosomes are called bivalents (two chromosomes) or tetrads (four chromatids), with one chromosome coming from each parent. Prophase I is divided into a series of substages which are named according to the appearance of chromosomes.

Leptotene

The first stage of prophase I is the *leptotene* stage, also known as *leptonema*, from Greek words meaning "thin threads". In this stage of prophase I, individual chromosomes—each consisting of two replicated sister chromatids—become "individualized" to form visible strands within the nucleus. The chromosomes each form a linear array of loops mediated by cohesin, and the lateral elements of the synaptonemal complex assemble forming an "axial element" from which the loops emanate. Recombination is initiated in this stage by the enzyme SPO11 which creates programmed double strand breaks (around 300 per meiosis in mice). This process generates single stranded DNA filaments coated by RAD51 and DMC1 which invade the homologous chromosomes, forming inter-axis bridges, and resulting in the pairing/co-alignment of homologues (to a distance of ~400 nm in mice).

Zygotene

Leptotene is followed by the *zygotene* stage, also known as *zygonema*, from Greek words meaning "paired threads", which in some organisms is also called the bouquet stage because of the way the telomeres cluster at one end of the nucleus. In this stage the homologous chromosomes become much more closely (~100 nm) and stably paired (a process called synapsis) mediated by the installation of the transverse and central elements of the synaptonemal complex. Synapsis is thought to occur in a zipper-like fashion starting from a recombination nodule. The paired chromosomes are called bivalent or tetrad chromosomes.

Pachytene

The *pachytene* stage (/┐ pækj┐ ti┐ n/ *PAK-i-teen*), also known as *pachynema*, from Greek words meaning "thick threads". is the stage at which all autosomal chromosomes have synapsed. In this stage homologous recombination, including chromosomal crossover (crossing over), is completed through the repair of the double strand breaks formed in leptotene. Most breaks are repaired without forming crossovers resulting in gene conversion. However, a subset of breaks (at least one per chromosome) form crossovers between non-sister (homologous) chromosomes resulting in the exchange of genetic information. Sex chromosomes, however, are not wholly identical, and only exchange information over a small region of homology called the pseudoautosomal region. The exchange of information between the homologous chromatids results in a recombination of information; each chromosome has the complete set of information it had before, and there are no gaps formed as a result of the process. Because the chromosomes cannot be distinguished in the synaptonemal complex, the actual act of crossing over is not perceivable through an ordinary light microscope, and chiasmata are not visible until the next stage.

Diplotene

During the *diplotene* stage, also known as *diplonema*, from Greek words meaning "two threads", the synaptonemal complex disassembles and homologous chromosomes separate from one another a little. However, the homologous chromosomes of each bivalent remain tightly bound at chiasmata, the regions where crossing-over occurred. The chiasmata remain on the chromosomes until they are severed at the transition to anaphase I to allow homologous chromosomes to move to opposite poles of the cell.

In human fetal oogenesis, all developing oocytes develop to this stage and are arrested in prophase I before birth. This suspended state is referred to as the *dictyotene stage* or dictyate. It lasts until meiosis is resumed to prepare the oocyte for ovulation, which happens at puberty or even later.

Diakinesis

Chromosomes condense further during the *diakinesis* stage, from Greek words meaning "moving through". This is the first point in meiosis where the four parts of the tetrads are actually visible. Sites of crossing over entangle together, effectively overlapping, making chiasmata clearly visible. Other than this observation, the rest of the stage closely resembles prometaphase of mitosis; the nucleoli disappear, the nuclear membrane disintegrates into vesicles, and the meiotic spindle begins to form.

Meiotic spindle formation

Unlike mitotic cells, human and mouse oocytes do not have centrosomes to produce the meiotic spindle. In mice, approximately 80 MicroTubule Organizing Centers (MTOCs) form a sphere in the ooplasm and begin to nucleate microtubules that reach out towards chromosomes, attaching to the chromosomes at the kinetochore. Over time the MTOCs merge until two poles have formed, generating a barrel shaped spindle. In human oocytes spindle microtubule nucleation begins on the chromosomes, forming an aster that eventually expands to surround the chromosomes. Chromosomes then slide along the microtubules towards the equator of the spindle, at which point the chromosome kinetochores form end-on attachments to microtubules.

Metaphase I

Homologous pairs move together along the metaphase plate: As *kinetochore microtubules* from both spindle poles attach to their respective kinetochores, the paired homologous chromosomes align along an equatorial plane that bisects the spindle, due to continuous counterbalancing forces exerted on the bivalents by the microtubules emanating from the two kinetochores of homologous chromosomes. This attachment is referred to as a bipolar attachment. The physical basis of the independent assortment of chromosomes is the random orientation of each bivalent along with the metaphase plate, with respect to the orientation of the other bivalents along the same equatorial line. The protein complex cohesin holds sister chromatids together from the time of their replication until anaphase. In mitosis, the force of kinetochore microtubules pulling in opposite directions creates tension. The cell senses this tension and does not progress with anaphase until all the chromosomes are properly bi-oriented. In meiosis, establishing tension ordinarily requires at least one crossover per chromosome pair in addition to cohesin between sister chromatids.

Anaphase I

Kinetochore microtubules shorten, pulling homologous chromosomes (which each consist of a pair of sister chromatids) to opposite poles. Nonkinetochore microtubules lengthen, pushing the centrosomes farther apart. The cell elongates in preparation for division down the center. Unlike in mitosis, only the cohesin from the chromosome arms is degraded while the cohesin surrounding the centromere remains protected by a protein named Shugoshin (Japanese for "guardian spirit"), what prevents the sister chromatids from separating. This allows the sister chromatids to remain together while homologs are segregated.

Telophase I

The first meiotic division effectively ends when the chromosomes arrive at the poles. Each daughter cell now has half the number of chromosomes but each chromosome consists of a pair of chromatids. The microtubules that make up the spindle network disappear, and a new nuclear membrane surrounds each haploid set. The chromosomes uncoil back into chromatin. Cytokinesis, the pinching of the cell membrane in animal cells or the formation of the cell wall in plant cells, occurs, completing the creation of two daughter cells. However, cytokinesis does not fully complete resulting in "cytoplasmic bridges" which enable the cytoplasm to be shared between daughter cells until the end of meiosis II. Sister chromatids remain attached during telophase I.

Cells may enter a period of rest known as interkinesis or interphase II. No DNA replication occurs during this stage.

Meiosis II

Meiosis II is the second meiotic division, and usually involves equational segregation, or separation of sister chromatids. Mechanically, the process is similar to mitosis, though its genetic results are fundamentally different. The end result is production of four haploid cells (n chromosomes; 23 in humans) from the two haploid cells (with n chromosomes, each consisting of two sister chromatids) produced in meiosis I. The four main steps of meiosis II are: prophase II, metaphase II, anaphase II, and telophase II.

In **prophase II**, we see the disappearance of the nucleoli and the nuclear envelope again as well as the shortening and thickening of the chromatids. Centrosomes move to the polar regions and arrange spindle fibers for the second meiotic division.

In **metaphase II**, the centromeres contain two kinetochores that attach to spindle fibers from the centrosomes at opposite poles. The new equatorial metaphase plate is rotated by 90 degrees when compared to meiosis I, perpendicular to the previous plate.

This is followed by **anaphase II**, in which the remaining centromeric cohesin, not protected by Shugoshin anymore, is cleaved, allowing the sister chromatids to segregate. The sister chromatids by convention are now called sister chromosomes as they move toward opposing poles.

The process ends with **telophase II**, which is similar to telophase I, and is marked by decondensation and lengthening of the chromosomes and the disassembly of the spindle. Nuclear envelopes re-form and cleavage or cell plate formation eventually produces a total of four daughter cells, each with a haploid set of chromosomes.

Meiosis is now complete and ends up with four new daughter cells.

Origin and function

The new combinations of DNA created during meiosis are a significant source of genetic variation alongside mutation, resulting in new combinations of alleles, which may be beneficial. Meiosis generates gamete genetic diversity in two ways: (1) Law of Independent Assortment. The independent orientation of homologous chromosome pairs along the metaphase plate during metaphase I and orientation of sister chromatids in metaphase II, this is the subsequent separation of homologs and sister chromatids during anaphase I and II, it allows a random and independent distribution of chromosomes to each daughter cell (and ultimately to gametes); and (2) Crossing Over. The physical exchange of homologous chromosomal regions by homologous recombination during prophase I results in new combinations of genetic information within chromosomes.

Prophase I arrest

Female mammals and birds are born possessing all the oocytes needed for future ovulations, and these oocytes are arrested at the prophase I stage of meiosis. In humans, as an example, oocytes are formed between three and four months of gestation within the fetus and are therefore present at birth. During this prophase I arrested stage (dictyate), which may last for decades, four copies of the genome are present in the oocytes. The arrest of ooctyes at the four genome copy stage was proposed to provide the informational redundancy needed to repair damage in the DNA of the germline. The repair process used appears to involve homologous recombinational repair Prophase I arrested oocytes have a high capability for efficient repair of DNA damage, particularly exogenously induced double-strand breaks. DNA repair capability appears to be a key quality control mechanism in the female germ line and a critical determinant of fertility.

Occurrence

In life cycles

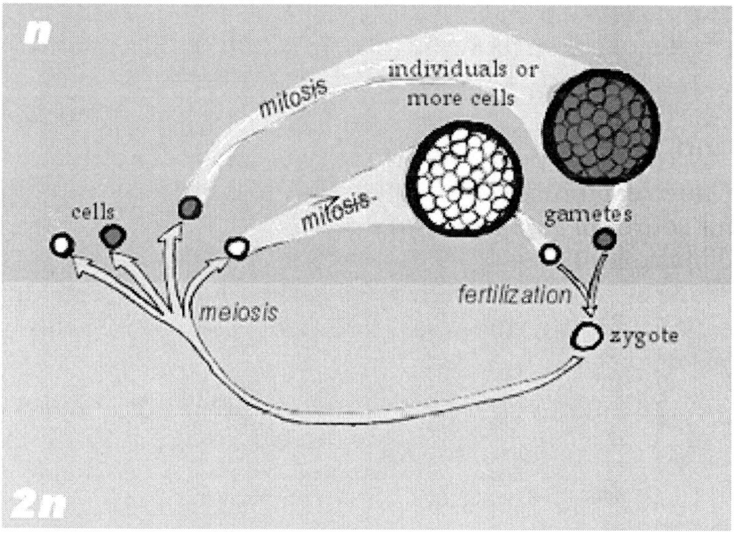

Diplontic life cycle

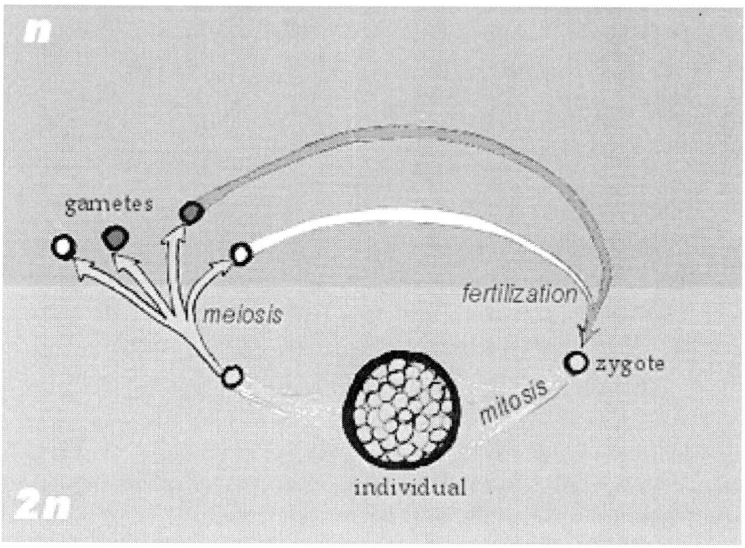

Haplontic life cycle.

Meiosis occurs in eukaryotic life cycles involving sexual reproduction, consisting of the constant cyclical process of meiosis and fertilization.

This takes place alongside normal mitotic cell division. In multicellular organisms, there is an intermediary step between the diploid and haploid transition where the organism grows. At certain stages of the life cycle, germ cells produce gametes. Somatic cells make up the body of the organism and are not involved in gamete production.

Cycling meiosis and fertilization events produces a series of transitions back and forth between alternating haploid and diploid states. The organism phase of the life cycle can occur either during the diploid state (*diplontic* life cycle), during the haploid state (*haplontic* life cycle), or both (*haplodiplontic* life cycle, in which there are two distinct organism phases, one during the haploid state and the other during the diploid state). In this sense there are three types of life cycles that utilize sexual reproduction, differentiated by the location of the organism phase(s).

In the *diplontic life cycle* (with pre-gametic meiosis), of which humans are a part, the organism is diploid, grown from a diploid cell called the zygote. The organism's diploid germ-line stem cells undergo meiosis to create haploid gametes (the spermatozoa for males and ova for females), which fertilize to form the zygote. The diploid zygote undergoes repeated cellular division by mitosis to grow into the organism.

In the *haplontic life cycle* (with post-zygotic meiosis), the organism is haploid instead, spawned by the proliferation and differentiation of a single haploid cell called the gamete. Two organisms of opposing sex contribute their haploid gametes to form a diploid zygote. The zygote undergoes meiosis immediately, creating four haploid cells. These cells undergo mitosis to create the organism. Many fungi and many protozoa utilize the haplontic life cycle.

Finally, in the *haplodiplontic life cycle* (with sporic or intermediate meiosis), the living organism alternates between haploid and diploid states. Consequently, this cycle is also known as the alternation of generations. The diploid organism's germ-line cells undergo meiosis to produce spores. The spores proliferate by mitosis, growing into a haploid organism. The haploid organism's gamete then combines with another haploid organism's gamete, creating the zygote. The zygote undergoes repeated mitosis and differentiation to become a diploid organism again. The haplodiplontic life cycle can be considered a fusion of the diplontic and haplontic life cycles.

In plants and animals

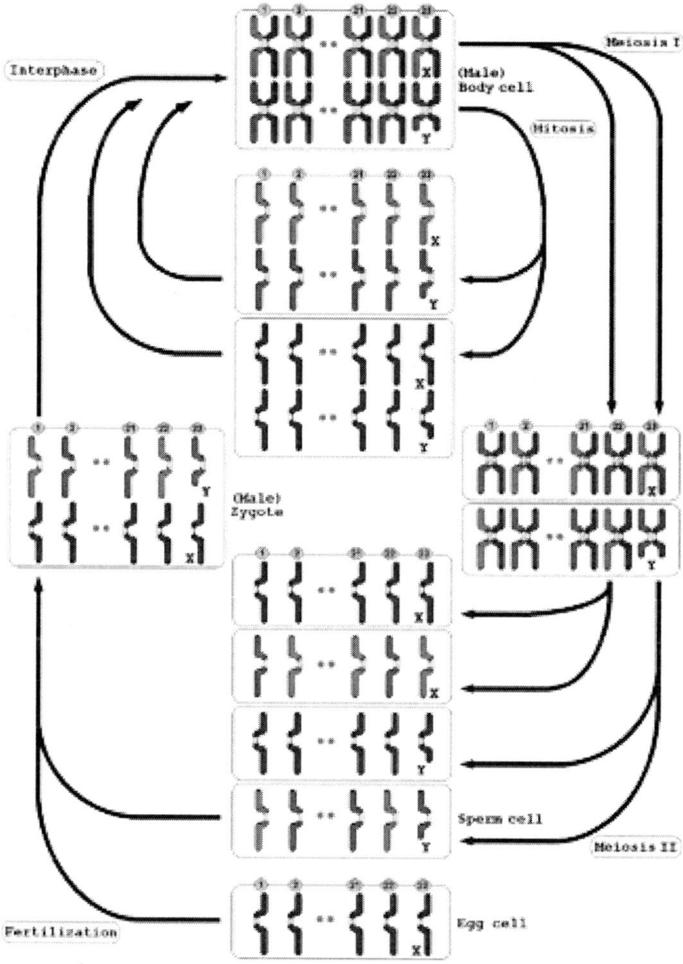

Overview of chromatides' and chromosomes' distribution within the mitotic and meiotic cycle of a male human cell

Meiosis occurs in all animals and plants. The end result, the production of gametes with half the number of chromosomes as the parent cell, is the same, but the detailed process is different. In animals, meiosis produces gametes directly. In land plants and some algae, there is an alternation of generations such that meiosis in the diploid sporophyte generation produces haploid spores. These spores multiply by mitosis, developing into the haploid gametophyte generation, which then gives rise to gametes directly (i.e. without further meiosis). In both animals and plants, the final stage is for the gametes to fuse, restoring the original number of chromosomes.

In mammals

In females, meiosis occurs in cells known as oocytes (singular: oocyte). Each primary oocyte divides twice in meiosis, unequally in each case. The first division produces a daughter cell, and a much smaller polar body which may or may not undergo a second division. In meiosis II, division of the daughter cell produces a second polar body, and a single haploid cell, which enlarges to become an ovum. Therefore, in females each primary oocyte that undergoes meiosis results in one mature ovum and one or two polar bodies.

Note that there are pauses during meiosis in females. Maturing oocytes are arrested in prophase I of meiosis I and lie dormant within a protective shell of somatic cells called the follicle. At the beginning of each menstrual cycle, FSH secretion from the anterior pituitary stimulates a few follicles to mature in a process known as folliculogenesis. During this process, the maturing oocytes resume meiosis and continue until metaphase II of meiosis II, where they are again arrested just before ovulation. If these oocytes are fertilized by sperm, they will resume and complete meiosis. During folliculogenesis in humans, usually one follicle becomes dominant while the others undergo atresia. The process of meiosis in females occurs during oogenesis, and differs from the typical meiosis in that it features a long period of meiotic arrest known as the dictyate stage and lacks the assistance of centrosomes.

In males, meiosis occurs during spermatogenesis in the seminiferous tubules of the testicles. Meiosis during spermatogenesis is specific to a type of cell called spermatocytes, which will later mature to become spermatozoa. Meiosis of primordial germ cells happens at the time of puberty, much later than in females. Tissues of the male testis suppress meiosis by degrading retinoic acid, proposed to be a stimulator of meiosis. This is overcome at puberty when cells within seminiferous tubules called Sertoli cells start making their own retinoic acid. Sensitivity to retinoic acid is also adjusted by proteins called nanos and DAZL. Genetic loss-of-function studies on retinoic acid-generating enzymes have shown that retinoic acid is required postnatally to stimulate spermatogonia differentiation which results several days later in spermatocytes undergoing meiosis, however retinoic acid is not required during the time when meiosis initiates.

In female mammals, meiosis begins immediately after primordial germ cells migrate to the ovary in the embryo. Some studies suggest that retinoic acid derived from the primitive kidney (mesonephros) stimulates meiosis in embryonic ovarian oogonia and that tissues of the embryonic male testis suppress meiosis by degrading retinoic acid. However, genetic loss-of-function studies on retinoic acid-generating enzymes have shown that retinoic acid is not required for initiation of either female meiosis which occurs during embryogenesis or male meiosis which initiates postnatally.

Flagellates

While the majority of eukaryotes have a two-divisional meiosis (though sometimes achiasmatic), a very rare form, one-divisional meiosis, occurs in some flagellates (parabasalids and oxymonads) from the gut of the wood-feeding cockroach *Cryptocercus*.

Role in human genetics and disease

Recombination among the 23 pairs of human chromosomes is responsible for redistributing not just the actual chromosomes, but also pieces of each of them. There is also an estimated 1.6-fold more recombination in females relative to males. In addition, average, female recombination is higher at the centromeres and male recombination is higher at the telomeres. On average, 1 million bp (1 Mb) correspond to 1 cMorgan (cm = 1% recombination frequency). The frequency of cross-overs remain uncertain. In yeast, mouse and human, it has been estimated that e"200 double-strand breaks (DSBs) are formed per meiotic cell. However, only a subset of DSBs (~5–30% depending on the organism), go on to produce crossovers, which would result in only 1-2 cross-overs per human chromosome.

Qualitative difference between sex chromosome & chromosome theory

Qualitative differences between chromosomes

The observations of T.H. Montgomery's (1901) and those of W.S. Sutton (1902) attained significance when T. **Boveri** (1902) demonstrated in sea urchin, that different chromosomes of a set possess different qualities. He also showed, that presence of a complete set of chromosomes was important for survival. This observation derives support now from the fact that a hypodiploid (chromosome number < *2n)* does not survive while a haploid individual may survive. It is also proved by the fact that trisomics involving different chromosomes differed morphologically as in case of *Datura* and many other plant species and also in Humans.

Sex-determination system

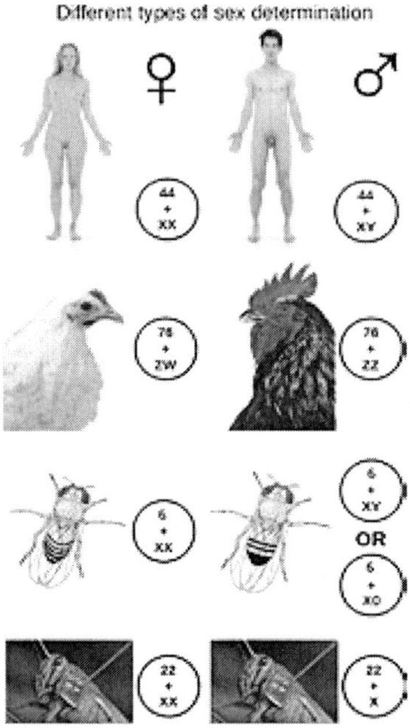

Some chromosomal sex determination systems in animals

A **sex-determination system** is a biological system that determines the development of sexual characteristics in an organism. Most organisms that create their offspring using sexual reproduction have two sexes.

In some species there are hermaphrodites. There are also some species that are only one sex due to parthenogenesis, the act of a female reproducing without fertilization.

In some species, sex determination is genetic: males and females have different alleles or even different genes that specify their sexual morphology. In animals this is often accompanied by chromosomal differences, generally through combinations of XY, ZW, XO, ZO chromosomes, or haplodiploidy. The sexual differentiation is generally triggered by a main gene (a "sex locus"), with a multitude of other genes following in a domino effect.

In other cases, sex of a fetus is determined by environmental variables (such as temperature). The details of some sex-determination systems are not yet fully understood. Hopes for future fetal biological system analysis include complete-reproduction-system initialized signals that can be measured during pregnancies to more accurately determine whether a determined sex of a fetus is male, or female. Such analysis of biological systems could also signal whether the fetus is hermaphrodite, which includes total or partial of both male and female reproduction organs.

Some species such as various plants and fish do not have a fixed sex, and instead go through life cycles and change sex based on genetic cues during corresponding life stages of their type. This could be due to environmental factors such as seasons and temperature. In some gonochoric species, a few individuals may have sex characteristics of both sexes, a condition called intersex.

While diversity in sex determination systems is common throughout the diversity of biology, the systems beyond XY/XX/XO in mammals are often characterized as less common or abnormal, because of the social influence of centering certain human cultural norms when researching sex determination. As such, general biology courses often focus on the simplified genetics of this one system, harmfully excluding people and other organisms from what most people learn in their biology courses.

Discovery

Sex determination was discovered in the mealworm by the American geneticist Nettie Stevens in 1903.

Chromosomal systems

XX/XY sex chromosomes

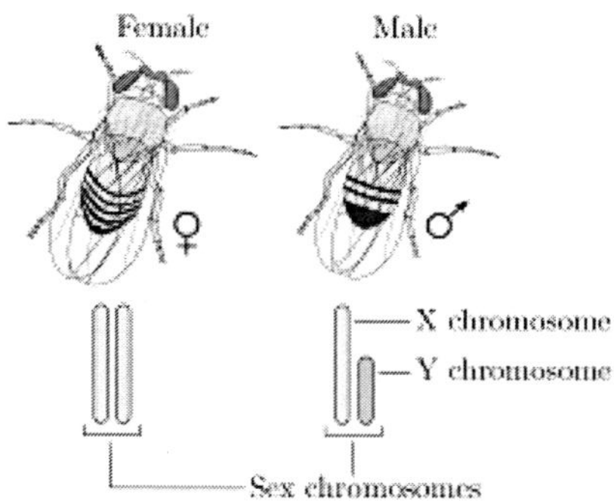

Drosophila sex-chromosomes

Human male XY chromosomes after G-banding

The **XX/XY sex-determination system** is the most familiar, as it is found in humans. The XX/XY system is found in most other mammals, as well as some insects. In this system, most females have two of the same kind of sex chromosome (XX), while most males have two distinct sex chromosomes (XY). The X and Y sex chromosomes are different in shape and size from each other, unlike the rest of the chromosomes (autosomes), and are sometimes called allosomes. In some species, such as humans, organisms remain sex indifferent for a time after they're created; in others, however, such as fruit flies, sexual differentiation occurs as soon as the egg is fertilized.

Y-centered sex determination

Some species (including humans) have a gene SRY on the Y chromosome that determines maleness. Members of SRY-reliant species can have uncommon XY chromosomal combinations such as XXY and still live. Human sex is determined by the presence or absence of a Y chromosome with a functional SRY gene. Once the SRY gene is activated, cells create testosterone and anti-müllerian hormone which typically ensures the development of a single, male reproductive system. In typical XX embryos, cells secrete estrogen, which drives the body toward the female pathway.

In Y-centered sex determination, the SRY gene is the main gene in determining male characteristics, but multiple genes are required to develop testes. In XY mice, lack of the gene DAX1 on the X chromosome results in sterility, but in humans it causes adrenal hypoplasia congenita. However, when an extra DAX1 gene is placed on the X chromosome, the result is a female, despite the existence of SRY. Even when there are normal sex chromosomes in XX females, duplication or expression of SOX9 causes testes to develop. Gradual sex reversal in developed mice can also occur when the gene FOXL2 is removed from females. Even though the gene DMRT1 is used by birds as their sex locus, species who have XY chromosomes also rely upon DMRT1, contained on chromosome 9, for sexual differentiation at some point in their formation.

X-centered sex determination

Some species, such as fruit flies, use the presence of two X chromosomes to determine femaleness. Species that use the number of Xs to determine sex are nonviable with an extra X chromosome.

Other variants of XX/XY sex determination

Some fish have variants of the XY sex-determination system, as well as the regular system. For example, while having an XY format, *Xiphophorus nezahualcoyotl* and *X. milleri* also have a second Y chromosome, known as Y', that creates XY' females and YY' males.

At least one monotreme, the platypus, presents a particular sex determination scheme that in some ways resembles that of the ZW sex chromosomes of birds and lacks the SRY gene. The platypus has ten sex chromosomes; males have an XYXYXYXYXY pattern while females have ten X chromosomes. Although it is an XY system, the platypus' sex chromosomes share no homologues with eutherian sex chromosomes. Instead, homologues with eutherian sex chromosomes lie on the platypus chromosome 6, which means that the eutherian sex chromosomes were autosomes at the time that the monotremes diverged from the therian mammals (marsupials and eutherian mammals). However, homologues to the avian DMRT1 gene on platypus sex chromosomes X3 and X5 suggest that it is possible the sex-determining gene for the platypus is the same one that is involved in bird sex-determination. More research must be conducted in order to determine the exact sex determining gene of the platypus.

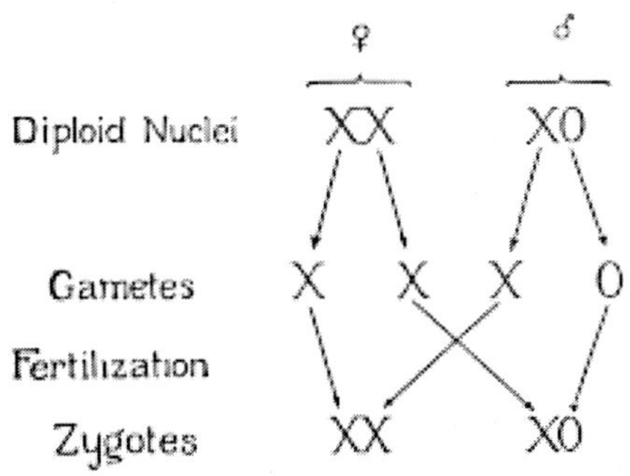

Heredity of sex chromosomes in XO sex determination

XX/X0 sex chromosomes

In this variant of the XY system, females have two copies of the sex chromosome (XX) but males have only one (X0). The *0* denotes the absence of a second sex chromosome. Generally in this method, the sex is determined by amount of genes expressed across the two chromosomes. This system is observed in a number of insects, including the grasshoppers and crickets of order Orthoptera and in cockroaches (order Blattodea). A small number of mammals also lack a Y chromosome. These include the Amami spiny rat (*Tokudaia osimensis*) and the Tokunoshima spiny rat (*Tokudaia tokunoshimensis*) and *Sorex araneus*, a shrew species. Transcaucasian mole voles (*Ellobius lutescens*) also have a form of XO determination, in which both sexes lack a second sex chromosome. The mechanism of sex determination is not yet understood.

The nematode *C. elegans* is male with one sex chromosome (X0); with a pair of chromosomes (XX) it is a hermaphrodite. Its main sex gene is XOL, which encodes XOL-1 and also controls the expression of the genes TRA-2 and HER-1. These genes reduce male gene activation and increase it, respectively.

ZW/ZZ sex chromosomes

The **ZW sex-determination system** is found in birds, some reptiles, and some insects and other organisms. The ZW sex-determination system is reversed compared to the XY system: females have two different kinds of chromosomes (ZW), and males have two of the same kind of chromosomes (ZZ). In the chicken, this was found to be dependent on the expression of DMRT1. In birds, the genes FET1 and ASW are found on the W chromosome for females, similar to how the Y chromosome contains SRY. However, not all species depend upon the W for their sex. For example, there are moths and butterflies that are ZW, but some have been found female with ZO, as well as female with ZZW. Also, while mammals deactivate one of their extra X chromosomes when female, it appears that in the case of Lepidoptera, the males produce double the normal amount of enzymes, due to having two Z's. Because the use of ZW sex determination is varied, it is still unknown how exactly most species determine their sex. However, reportedly, the silkworm *Bombyx mori* uses a single female-specific piRNA as the primary determiner of sex. Despite the similarities between the ZW and XY systems, these sex chromosomes evolved separately. In the case of the chicken, their Z chromosome is more similar to humans' autosome 9. The chicken's Z chromosome also seems to be related to the X chromosome of the platypus. When a ZW species, such as the Komodo dragon, reproduces parthenogenetically, usually only males are produced. This is due to the fact that the haploid eggs double their chromosomes, resulting in ZZ or WW. The ZZ become males, but the WW are not viable and are not brought to term.

In both XY and ZW sex determination systems, the sex chromosome carrying the critical factors is often significantly smaller, carrying little more than the genes necessary for triggering the development of a given sex.

ZZ/Z0 sex chromosomes

The **ZZ/Z0 sex-determination system** is found in some moths. In these insects there is one sex chromosome, Z. Males have two Z chromosomes, whereas females have one Z. Males are ZZ, while females are Z0.

UV sex chromosomes

In some Bryophyte and some algae species, the gametophyte stage of the life cycle, rather than being hermaphrodite, occurs as separate male or female individuals that produce male and female gametes respectively. When meiosis occurs in the sporophyte generation of the life cycle, the sex chromosomes known as U and V assort in spores that carry either the U chromosome and give rise to female gametophytes, or the V chromosome and give rise to male gametophytes.

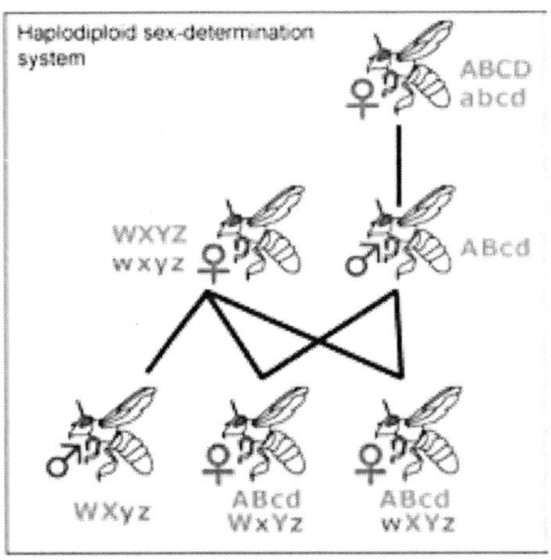

Haplodiploid sex chromosomes

Haplodiploidy

Haplodiploidy is found in insects belonging to Hymenoptera, such as ants and bees. Sex determination is controlled by the zygosity of a complementary sex determiner (*csd*) locus. Unfertilized eggs develop into haploid individuals which have a single, hemizygous copy of the *csd* locus and are therefore males. Fertilized eggs develop into diploid individuals which, due to high variability in the *csd* locus, are generally heterozygous females. In rare instances diploid individuals may be homozygous, these develop into sterile males. The gene acting as a *csd* locus has been identified in the honeybee and several candidate genes have been proposed as a *csd* locus for other Hymenopterans. Most females in the Hymenoptera order can decide the sex of their offspring by holding received sperm in their spermatheca and either releasing it into their oviduct or not. This allows them to create more workers, depending on the status of the colony.

Other chromosomal systems

Other uncommon systems include those of the green swordtail (a polyfactorial system with the sex-determining genes on several chromosomes) the Chironomus midges; the juvenile hermaphroditism of zebrafish, with an unknown trigger; and the platyfish, which has W, X, and Y chromosomes. This allows WY, WX, or XX females and YY or XY males.

Mating type in microorganisms is analogous to sex in multi-cellular organisms, and is sometimes described using those terms, though they are not necessarily correlated with physical body structures. Some species have more than two mating types. *Tetrahymena,* a type of ciliate, has seven mating types. *Schizophyllum commune,* a type of fungus, has 23,328.

Environmental systems

All alligators determine the sex of their offspring by the temperature of the nest.

Temperature-dependent

Many other sex-determination systems exist. In some species of reptiles, including alligators, some turtles, and the tuatara, sex is determined by the temperature at which the egg is incubated during a temperature-sensitive period. There are no examples of temperature-dependent sex determination (TSD) in birds. Megapodes had formerly been thought to exhibit this phenomenon, but were found to actually have different temperature-dependent embryo mortality rates for each sex. For some species with TSD, sex determination is achieved by exposure to hotter temperatures resulting in the offspring being one sex and cooler temperatures resulting in the other. This type of TSD is called *Pattern I*. For others species using TSD, it is exposure to temperatures on both extremes that results in offspring of one sex, and exposure to moderate temperatures that results in offspring of the opposite sex, called *Pattern II* TSD. The specific temperatures required to produce each sex are known as the female-promoting temperature and the male-promoting temperature. When the temperature stays near the threshold during the temperature sensitive period, the sex ratio is varied between the two sexes. Some species' temperature standards are based on when a particular enzyme is created. These species that rely upon temperature for their sex determination do not have the SRY gene, but have other genes such as DAX1, DMRT1, and SOX9 that are expressed or not expressed depending on the temperature. The sex of some species, such as the Nile tilapia, Australian skink lizard, and Australian dragon lizard, has an initial bias, set by chromosomes, but can later be changed by the temperature of incubation.

It is unknown how exactly temperature-dependent sex determination evolved. It could have evolved through certain sexes being more suited to certain areas that fit the temperature requirements. For example, a warmer area could be more suitable for nesting, so more females are produced to increase the amount that nest next season. In amniotes, environmental sex determination preceded the genetically determined systems of birds and mammals; it is thought that a temperature-dependent amniote was the common ancestor of amniotes with sex chromosomes.

Other environmental systems

There are other environmental sex determination systems including location-dependent determination systems as seen in the marine worm *Bonellia viridis* – larvae become males if they make physical contact with a female, and females if they end up on the bare sea floor. This is triggered by the presence of a chemical produced by the females, bonellin. Some species, such as some snails, practice sex change: adults start out male, then become female. In tropical clown fish, the dominant individual in a group becomes female while the other ones are male, and bluehead wrasses (*Thalassoma bifasciatum*) are the reverse. Some species, however, have no sex-determination system. Hermaphrodite species include the common earthworm and certain species of snails. A few species of fish, reptiles, and insects reproduce by parthenogenesis and are female altogether. There are some reptiles, such as the boa constrictor and Komodo dragon that can reproduce both sexually and asexually, depending on whether a mate is available.

Evolution

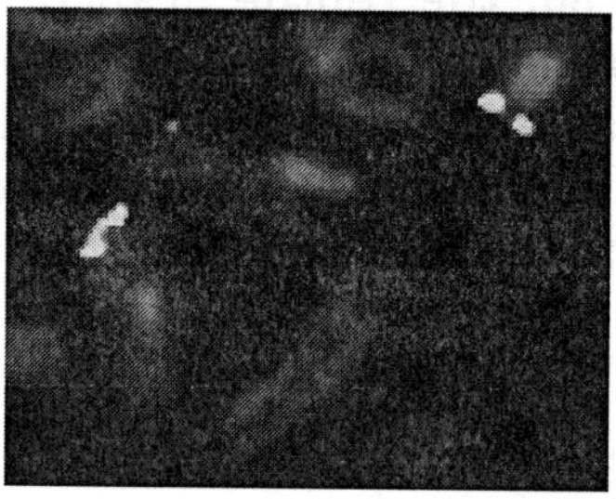

The ends of the XY chromosomes, highlighted here in green, are all that is left of the original autosomes that can still cross over with each other.

Sex determination systems may have evolved from mating type, which is a feature of microorganisms.

Chromosomal sex determination may have evolved early in the history of eukaryotes. But in plants it has been suggested to have evolved recently.

The accepted hypothesis of XY and ZW sex chromosome evolution in amniotes is that they evolved at the same time, in two different branches.

No genes are shared between the avian ZW and mammal XY chromosomes and the chicken Z chromosome is similar to the human autosomal chromosome 9, rather than X or Y. This suggests not that the ZW and XY sex-determination systems share an origin but that the sex chromosomes are derived from autosomal chromosomes of the common ancestor of birds and mammals. In the platypus, a monotreme, the X_1 chromosome shares homology with therian mammals, while the X_5 chromosome contains an avian sex-determination gene, further suggesting an evolutionary link.

However, there is some evidence to suggest that there could have been transitions between ZW and XY, such as in *Xiphophorus maculatus*, which have both ZW and XY systems in the same population, despite the fact that ZW and XY have different gene locations. A recent theoretical model raises the possibility of both transitions between the XY/XX and ZZ/ZW system and environmental sex determination The platypus' genes also back up the possible evolutionary link between XY and ZW, because they have the DMRT1 gene possessed by birds on their X chromosomes. Regardless, XY and ZW follow a similar route. All sex chromosomes started out as an original autosome of an original amniote that relied upon temperature to determine the sex of offspring. After the mammals separated, the reptile branch further split into Lepidosauria and Archosauromorpha. These two groups both evolved the ZW system separately, as evidenced by the existence of different sex chromosomal locations. In mammals, one of the autosome pair, now Y, mutated its SOX3 gene into the SRY gene, causing that chromosome to designate sex. After this mutation, the SRY-containing chromosome inverted and was no longer completely homologous with its partner. The regions of the X and Y chromosomes that are still homologous to one another are known as the pseudoautosomal region. Once it inverted, the Y chromosome became unable to remedy deleterious mutations, and thus degenerated. There is some concern that the Y chromosome will shrink further and stop functioning in ten million years: but the Y chromosome has been strictly conserved after its initial rapid gene loss.

There are some vertebrate species, such as the medaka fish, that evolved sex chromosomes separately; their Y chromosome never inverted and can still swap genes with the X. These species' sex chromosomes are relatively primitive and unspecialized. Because the Y does not have male-specific genes and can interact with the X, XY and YY females can be formed as well as XX males. Non-inverted Y chromosomes with long histories are found in pythons and emus, each system being more than 120 million years old, suggesting that inversions are not necessarily an eventuality. XO sex determination can evolve from XY sex determination with about 2 million years.

Sex chromosome

A **sex chromosome** (also referred to as an **allosome**, **heterotypical chromosome, gonosome**, **heterochromosome**, or **idiochromosome**) is a chromosome that differs from an ordinary autosome in form, size, and behavior. The human sex chromosomes, a typical pair of mammal allosomes, determine the sex of an individual created in sexual reproduction. Autosomes differ from allosomes because autosomes appear in pairs whose members have the same form but differ from other pairs in a diploid cell, whereas members of an allosome pair may differ from one another and thereby determine sex.

Nettie Stevens and Edmund Beecher Wilson both independently discovered sex chromosomes in 1905. However, Stevens is credited for discovering them earlier than Wilson.

Differentiation

In humans, each cell nucleus contains 23 pairs of chromosomes, a total of 46 chromosomes. The first 22 pairs are called autosomes. Autosomes are homologous chromosomes i.e. chromosomes which contain the same genes (regions of DNA) in the same order along their chromosomal arms. The 23rd pair of chromosomes are called allosomes. These consist of two X chromosomes in most females, and an X chromosome and a Y chromosome in most males. Females therefore have 23 homologous chromosome pairs, while males have 22. The X and Y chromosomes have small regions of homology called pseudoautosomal regions.

An X chromosome is always present as the 23rd chromosome in the ovum, while either an X or Y chromosome may be present in an individual sperm. Early in female embryonic development, in cells other than egg cells, one of the X chromosomes is randomly and permanently partially deactivated: In some cells, the X chromosome inherited from the mother deactivates; in other cells, it is the X chromosome inherited from the father. This ensures that both sexes always have exactly one functional copy of an X chromosome in each body cell. The deactivated X chromosome is silenced by repressive heterochromatin that compacts the DNA and prevents expression of most genes. This compaction is regulated by PRC2 (Polycomb Repressive Complex 2).

Sex determination

All diploid organisms with allosome-determined sex get half of their allosomes from each of their parents. In most mammals, females are XX, and can pass along either of their Xs; since males are XY they can pass along either an X or a Y. Females in such species receive an X chromosome from each parent while males receive an X chromosome from their mother and a Y chromosome from their father. It is thus the male's sperm that determines the sex of each offspring in such species.

However, a small percentage of humans have a divergent sexual development, known as intersex. This can result from allosomes that are neither XX nor XY. It can also occur when two fertilized embryo fuse, producing a chimera that might contain two different sets of DNA one XX and the other XY. It could also result from exposure, often in utero, to chemicals that disrupt the normal conversion of the allosomes

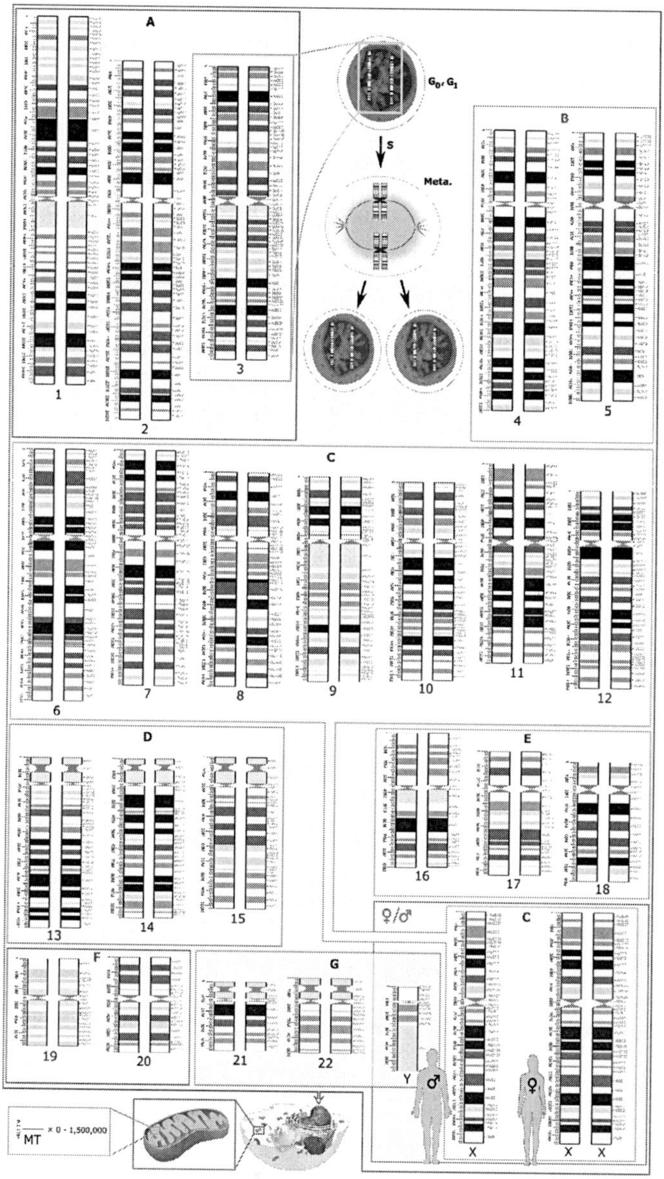

Figure: Schematic karyogram of a human, showing the sex chromosomes in green box at bottom right. The X chromosome is part of chromosome group C, and the Y chromosome is part of group G. Bands and sub-bands are annotated to the right of each chromosome (or chromosome pair), and the gene for the sex-determining region Y protein is located at Yp11.2.

into sex hormones and further into the development of either ambiguous outer genitalia or internal organs.

There is a gene in the Y chromosome that has regulatory sequences that control genes that code for maleness, called the SRY gene. This gene produces a testis-determining factor ("TDF"), which initiates testis development in humans and other mammals. The SRY sequence's prominence in sex determination was discovered when the genetics of sex-reversed XX men (i.e. humans who possess biological male-traits but actually have XX allosomes) were studied. After examination, it was discovered that the difference between a typical XX individual (traditional female) and a sex-reversed XX man was that the typical individuals lacked the SRY gene. It is theorized that in sex-reversed XX men, the SRY mistakenly gets translocated to an X chromosome in the XX pair during meiosis.

Other vertebrates

Diverse mechanisms are involved in the determination of sex in animals. For mammals, sex determination is carried by the genetic contribution of the spermatozoon. Many lower chordates, such as fish, amphibians and reptiles, have systems that are influenced by the environment. Fish and amphibians, for example, have genetic sex determination but their sex can also be influenced by externally available steroids and incubation temperature of eggs. In some reptiles, e.g. sea turtles, only the incubation temperature determines sex (temperature-dependent sex determination).

Plants

Many scientists argue that sex determination in plants is more complex than that in humans. This is because even flowering plants have a variety of mating systems, their sex determination primarily regulated by MADS-box genes. These genes code for proteins that form the sex organs in flowers.

Plant sex chromosomes are most common in bryophytes, relatively common in vascular plants and unknown in ferns and lycophytes. The diversity of plants is reflected in their sex-determination systems, which include XY and UV systems as well as many variants. Sex chromosomes have evolved independently across many plant groups. Recombination of chromosomes may lead to heterogamety before the development of sex chromosomes, or recombination may be reduced after sex chromosomes develop. Only a few pseudoautosomal regions normally remain once sex chromosomes are fully differentiated. When chromosomes do not recombine, neutral sequence divergences begin to accumulate, which has been used to estimate the age of sex chromosomes in various plant lineages. Even the oldest estimated divergence, in the liverwort *Marchantia polymorpha,* is more recent than mammal or bird divergence. Due to this recency, most plant sex chromosomes also have relatively small sex-linked regions. Current evidence does not support the existence of plant sex chromosomes more ancient than those of *M. polymorpha.*

The high prevalence of autopolyploidy in plants also impacts the structure of their sex chromosomes. Polyploidization can occur before and after the development of sex chromosomes. If it occurs after sex chromosomes are established, dosage should stay consistent between the sex chromosomes and autosomes, with minimal impact on sex differentiation. If it occurs before sex chromosomes become heteromorphic, as is likely in the octoploid red sorrel *Rumex acetosella,* sex is determined in a single XY system. In a more complicated system, the sandalwood species *Viscum fischeri* has X1X1X2X2 chromosomes in females, and X1X2Y chromosomes in males.

Sequence composition and evolution

Amplification of transposable elements, tandom repeats especially accumulation of long tandom repeats (LTR) retrotransposones are responsible for plant sex chromosome evolution. The insertion of retrotransposons is probably the major cause of y-chromosome expansion and plant genome size evolution. Retrotransposones contribute in size determination of sex chromosomes and its proliferation varies even in closely related species. LTR and tandom repeats play dominant role in the evolution of *S. latifolia* sex chromosomes. Athila is new family of retroelements, discovered in *Arabidopsis thaliana,* present in heterochromatin region only. Athila retroelements overrepresented in X but absent in Y while tandem repeats enriched in Y-chromosome. Some chloroplast sequences have also been identified in the Y-chromosome of *S. latifolia. S. vulgaris* has more retroelements in their sex chromosomes compare to *S. latifolia.* Microsatellite data shows that there is no significant difference between X and Y-chromosome microsatellites in both *Silene* species. This would conclude that microsatellites do not participate in Y-chromosome evolution. The portion of Y-chromosome that never recombine with X-chromosome faces selection reduction. This reduced selection leads to insertion of transposable elements and accumulation of deleterious mutation. The Y become larger and smaller than X due to insertion of retroelement and deletion of genetic material respectively. The genus *Humulus* is also used as model for the study of sex chromosomes evolution. Based on the phylogenetic topology distribution there are three regions on sex chromosomes. One region that stops recombining in the ancestor of *H. lupulus*, second that stops recombining in modern *H. lupulus* and the third region called pseudoautosomal region. *H. lupulus* is the rare case in plants in which Y is smaller than X, while its ancestor plant has the same size of both X and Y chromosomes. This size difference should be caused by deletion of genetic material in Y but that is not the case. This is because of complex dynamics like the larger size of X than Y-chromosome may be due to duplication or retrotransposition and size of Y remains same.

Non-vascular plants

Ferns and lycophytes have bisexual gametophytes, so there is no evidence for sex chromosomes. In the bryophytes, including liverworts, hornworts and mosses, sex chromosomes are common. The sex chromosomes in bryophytes affect what type of gamete is produced by the gametophyte, and there is wide diversity in gametophyte type. Unlike seed plants, where gametophytes are always unisexual, in bryophytes they may produce male, female, or both types of gamete.

Bryophytes most commonly employ a UV sex-determination system, where U produces female gametophytes and V produces male gametophytes. The U and V chromosomes are heteromorphic with U larger than V and are frequently both larger than the autosomes. There is variation even within this system, including UU/V and U/VV chromosome arrangements. In some bryophytes, microchromosomes have been found to co-occur with sex chromosomes and likely impact sex determination.

Gymnosperms

Dioecy is common among gymnosperms, found in an estimated 36% of species. However, heteromorphic sex chromosomes are relatively rare, with only five species known as of 2014. Five of these use an XY system, and one (*Ginkgo biloba*) uses a WZ system. Some gymnosperms, such as Johann's Pine (*Pinus johannis*), have homomorphic sex chromosomes that are almost indistinguishable through karyotyping.

Angiosperms

Cosexual angiosperms with either monoecious or hermaphroditic flowers do not have sex chromosomes. Angiosperms with separate sexes (dioecious) may use sex chromosomes or environmental flowers for sex determination. Cytogenetic data from about 100 angiosperm species showed heteromorphic sex chromosomes in approximately half, mostly taking the form of XY sex-determination systems. Their Y is typically larger, unlike in humans; however there is diversity among angiosperms. In the Poplar genus (*Populus*) some species have male heterogamety while others have female heterogamety. Sex chromosomes have arisen independently multiple times in angiosperms, from the monoecious ancestral condition. The move from a monoecious to dioecious system requires both male and female sterility mutations to be present in the population. Male sterility likely arises first as an adaptation to prevent selfing. Once male sterility has reached a certain prevalence, then female sterility may have a chance to arise and spread.

In the domesticated papaya (*Carica papaya*), three sex chromosomes are present, denoted as X, Y and Y^h. This corresponds with three sexes: females with XX chromosomes, males with XY, and hermaphrodites with XY^h. The hermaphrodite sex is estimated to have arisen only 4000 years ago, post-domestication of the plant. The genetic architecture suggests that either the Y chromosome has an X-inactivating gene, or that the Y^h chromosome has an X-activating gene.

Medical applications

Allosomes not only carry the genes that determine male and female traits, but also those for some other characteristics as well. Genes that are carried by either sex chromosome are said to be sex linked. Sex linked diseases are passed down through families through one of the X or Y chromosomes. Since usually men inherit Y chromosomes, they are the only ones to inherit Y-linked traits. Men and women can get the X-linked ones since both inherit X chromosomes.

An allele is either said to be dominant or recessive. Dominant inheritance occurs when an abnormal gene from one parent causes disease even though the matching gene from the other parent is normal. The abnormal allele dominates. Recessive inheritance is when both matching genes must be abnormal to cause disease. If only one gene in the pair is abnormal, the disease does not occur, or is mild. Someone who has one abnormal gene (but no symptoms) is called a carrier. A carrier can pass this abnormal gene to his or her children. X chromosome carry about 1500 genes, more than any other chromosome in the human body. Most of them code for something other than female anatomical traits. Many of the non-sex determining X-linked genes are responsible for abnormal conditions. The Y chromosome carries about 78 genes. Most of the Y chromosome genes are involved with essential cell house-keeping activities and sperm production. Only one of the Y chromosome genes, the SRY gene, is responsible for male anatomical traits. When any of the 9 genes involved in sperm production are missing or defective the result is usually very low sperm counts and infertility. Examples of mutations on the X chromosome include more common diseases such as the following:

- Color blindness or color vision deficiency is the inability or decreased ability to see color, or perceive color differences, under normal lighting conditions. Color blindness affects many individuals in the population. There is no actual blindness, but there is a deficiency of color vision. The most usual cause is a fault in the development of one or more sets of retinal cones that perceive color in light and transmit that information to the optic nerve. This type of color blindness is usually a sex-linked condition. The genes that produce photopigments are carried on the X chromosome; if some of these genes are missing or damaged, color blindness will be expressed in males with a higher probability than in females because males only have one X chromosome.
- Hemophilia refers to a group of bleeding disorders in which it takes a long time for the blood to clot. This is referred to as X-Linked recessive. Hemophilia is much more common in males than females because males are hemizygous. They only have one copy of the gene in question and therefore express the trait when they inherit one mutant allele. In contrast, a female must inherit two mutant alleles, a less frequent event since the mutant allele is rare in the population. X-linked traits are maternally inherited from carrier mothers or from an affected father. Each son born to a carrier mother has a 50% probability of inheriting the X chromosome carrying the mutant allele.
- Fragile X syndrome is a genetic condition involving changes in part of the X chromosome. It is the most common form of inherited intellectual disability (mental retardation) in males. It is caused by a change in a gene called FMR1. A small part of the gene code is repeated on a fragile area of the X chromosome. The more repeats, the more likely there is to be a problem. Males and females can both be affected, but because males have only one X chromosome, a single fragile X is likely to affect them more. Most fragile-X males have large testes, big ears, narrow

faces, and sensory processing disorders that result in learning disabilities.

Other complications include:

- 46,XX testicular disorder of sex development, also called XX male syndrome, is a condition in which individuals with two X chromosomes in each cell, the pattern normally found in females, have a male appearance. People with this disorder have male external genitalia. In most people with 46,XX testicular disorder of sex development, the condition results from an exchange of genetic material between chromosomes (translocation). This exchange occurs as a random event during the formation of sperm cells in the affected person's father. The SRY gene (normally on the Y chromosome) is misplaced in this disorder, onto an X chromosome. Any person with an X chromosome that carries the SRY gene will develop male characteristics despite not having a Y chromosome.

Evolution

Sex chromosomes evolve from standard pairs of autosomal chromosomes. In a large number of organisms, the sex-determination systems presently observed are products of sex chromosome turnover. Sex chromosome turnover is a process defined as when the type of the sex chromosome changes as a product of a change in the identity of the sex-determining genes (such as by mutation) or by a change in their location. In other cases, sex chromosomes may grow substantially with respect to their ancestral forms as a result of fusion events with autosomes, and autosome-sex chromosome fusions result in what are called neo-sex chromosomes. Five examples of this are now known in the songbird superfamily *Sylvioidea.* There is one experimentally documented case of sex chromosome turnover occurring during a 30-year evolutionary experiment involving teleost fish, in which hybridization experiments resulted in a translocation of the sex-determiner region of a sex chromosome into an autosome. This resulted in the autosome becoming a novel W sex chromosome.

Developing the Chromosome Theory

The second half of the nineteenth century was a time of remarkable advances in genetic research. As early as the 1860s, Gregor Mendel and Charles Darwin began to explore possible mechanisms of heredity. Then, over the next few decades, Walther Flemming, Theodor Boveri, and Walter Sutton made a series of significant discoveries involving chromosomes, including what these structures looked like, how they moved during mitosis, and what role they likely played in the transmission of genetic characteristics. However, it wasn't until the work of Thomas Hunt Morgan in the early twentieth century that researchers were finally able to directly link the inheritance of genetic traits to the behavior of chromosomes, thereby providing concrete evidence for what became known as the chromosome theory of heredity.

Early Breakthroughs: Mendel and Darwin Propose Mechanisms of Heredity

The year 1865 was marked by two profound biological breakthroughs: Mendel's publication of his experiments in plant hybridization, and Darwin's provisional hypothesis of pangenesis. Although neither man had any direct evidence for the theories he proposed, Mendel speculated that cells contained some type of factor that carried traits from one generation to the next, while Darwin proposed that traits could be passed down via units he termed "gemmules," which he believed traveled from every body part to the sexual organs, where they were stored (Benson, 2001). Because these first attempts to explain the mechanisms of heredity lacked any scientific support, their profound importance went unrecognized by the scientific community for decades. Nonetheless, Mendel and Darwin's work laid the foundation for formulating a testable, research-based theory of heredity.

Flemming, Boveri, and Sutton Connect Chromosomes to Heredity

Meanwhile, several independent advances-namely, discovery of chromosomes and their behavior during mitosis as described by , and discovery of the link between chromosomes and heredity as described by Theodor Boveri-began to shed light on the events that take place in the nucleus during cell division and their implications. Then, some thirty-five years later, the significance of Mendel's work was emphasized by Walter Sutton, whose observations of chromosome behavior during cell division and gamete formation were consistent with Mendel's findings. Thus, the basis for chromosome theory, and the field of cytogenetics, was created.

Walther Flemming Describes Chromosomes

By the middle of the nineteenth century, scientists understood that cells derived from other cells and that the hereditary information was located in the nucleus, but the physical nature of the hereditary material

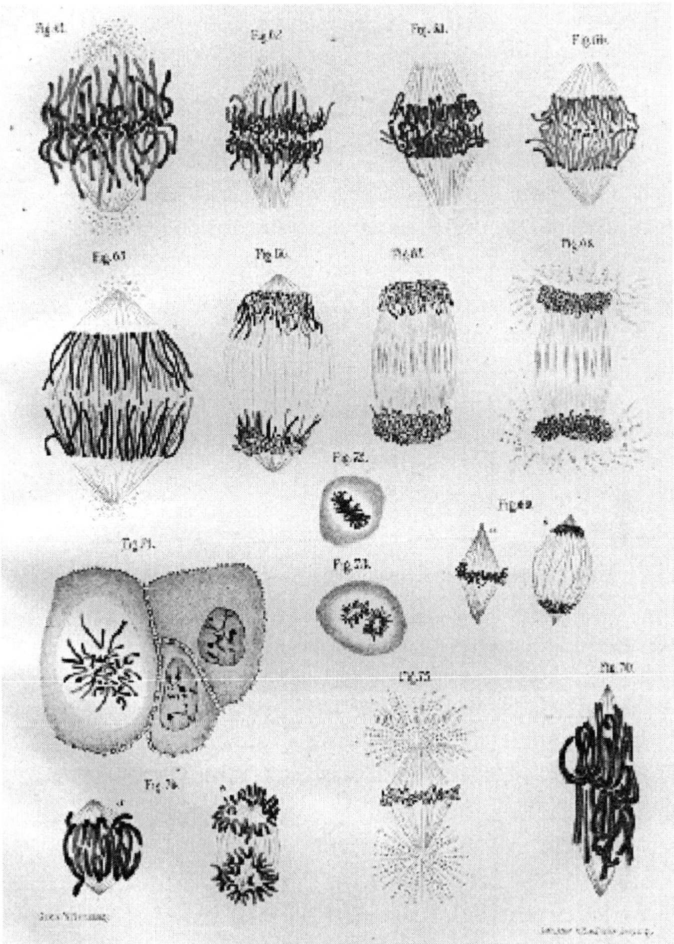

Figure : Drawing of mitosis by Walther Flemming.

remained unknown. The microscopes of the time provided very poor resolution of living cellular structures, making it necessary for investigators to treat fixed cells with various stains to enhance the contrast of their contents. Thus, using innovative microscopy techniques and painstaking precision, German anatomist Walther Flemming recognized and explored the fibrous network within the nucleus, which he termed chromatin, or "stainable material." (Flemming had actually discovered the chromosome, although the term would be coined a few years later by Heinrich Waldeyer.) Flemming noted that during cell division, the chromatin formed threadlike bodies, which he termed mitosen, from the Greek word for thread. Based on many observations of cells in various stages of division, Flemming correctly deduced the sequence of chromosome movements during mitosis, which would be confirmed decades later by microscopy of live dividing cells. With his characteristic attention to detail, Flemming also made the important observation that chromosomes split along their length during mitosis, and he correctly hypothesized that the split chromosomes were partitioned into different daughter cells at the end of mitosis.

Thus, Flemming recognized that chromosomal movement during mitosis offered a mechanism for the precise distribution of nuclear material during cell division. His work provided an invaluable description of the initial mechanisms underlying the process of cell division, and it helped paved the way for the discovery of hereditary mechanisms.

Theodor Boveri Links Chromosomes and Heredity

The end of the nineteenth century was marked by advancements in cytological techniques and microscopy. During this period, German embryologist Theodor Boveri took Flemming's findings to the next level by providing the first evidence that chromosomes of germ cell lineage provide continuity between generations. Boveri found evidence for this hypothesis through his research of early development in the roundworm species *Ascaris megalocephala* (now known as *Parascaris equorum*). *Ascaris* embryos provided an excellent experimental model for Boveri's observations; this is because the large, clear cells of the *Ascaris* embryo have only two pairs of chromosomes, and the embryos develop distinct somatic cell and germ cell lineages during the first few cleavage divisions. Boveri was thus able to trace the fate of the chromosomes in individual cell lineages with great precision, and he made the surprising observation that the full complement of chromosomes was retained only in the *Ascaris* germ cell lineage, the source of future gametes. By contrast, the chromosomes in somatic cells, the source of all other adult tissues, underwent a curious process of fragmentation and elimination known as chromosome diminution. (Chromosome diminution does not occur in mammals.)

Boveri also recognized that chromosome number was reduced in the gametes. In *Ascaris* eggs, meiosis does not occur until fertilization is complete, so Boveri was able to observe the behavior of egg and sperm chromosomes following fertilization. He noted that *Ascaris* eggs retained only two chromosomes after the polar body formed, and that the normal number of four chromosomes was restored following fusion of the sperm and egg pronuclei. Boveri's work provided one of the first descriptions of meiosis.

Walter Sutton Finds Evidence for Mendel's Principles

Working independently near the dawn of the twentieth century, American graduate student Walter Sutton confirmed and expanded upon Boveri's observations using a superior cytological model. Sutton, a Kansas farm boy, had made the serendipitous discovery that it was possible to distinguish individual chromosomes in cells undergoing meiosis in the testes of the lubber grasshopper, *Brachystola magna*. Thus, in his classic paper published in 1902, Sutton described the configurations of individual chromosomes in cells at various stages of meiosis. Sutton was able to identify 11 pairs of chromosomes that could be distinguished by their sizes, as well as an accessory singleton that he correctly presumed to be a sex chromosome. Due to its apparent lack of a partner, this chromosome sorted into only half of the sperm cells, while the remaining half never received a copy. Though the accessory chromosome was unpaired, it still replicated and entered stages of mitosis in the same manner as all other chromosomes, which prompted Sutton to declare it a "true chromosome" and not simply an accessory.

Sutton postulated that all chromosomes have a stable structure, or "individuality," that is maintained between generations, and he used this property to follow the behavior of individual chromosomes through the various stages of meiosis, including synapsis. Most notably, Sutton recognized that his observations were consistent with the , whose findings had only recently been rediscovered. In fact, Sutton closed his 1902 paper with the statement, "I may finally call attention to the probability that the association of paternal and maternal chromosomes in pairs and their subsequent separation during the reducing division as indicated above may constitute the physical basis of the Mendelian law of heredity." With these words, Sutton articulated the chromosomal theory of inheritance.

Chapter 6

Cell : Division

Cell division is the process by which a parent cell divides into two daughter cells. Cell division usually occurs as part of a larger cell cycle in which the cell grows and replicates its chromosome(s) before dividing. In eukaryotes, there are two distinct types of cell division: a vegetative division (mitosis), producing daughter cells genetically identical to the parent cell, and a cell division that produces haploid gametes for sexual reproduction (meiosis), reducing the number of chromosomes from two of each type in the diploid parent cell to one of each type in the daughter cells. In cell biology, **mitosis** (/majˈ toˈ sjˈ s/) is a part of the cell cycle, in which, replicated chromosomes are separated into two new nuclei. Cell division gives rise to genetically identical cells in which the total number of chromosomes is maintained. In general, mitosis (division of the nucleus) is preceded by the S stage of interphase (during which the DNA replication occurs) and is often followed by telophase and cytokinesis; which divides the cytoplasm, organelles, and cell membrane of one cell into two new cells containing roughly equal shares of these cellular components. The different stages of mitosis all together define the **mitotic (M) phase** of animal cell cycle—the division of the mother cell into two genetically identical daughter cells. Meiosis results in four haploid daughter cells by undergoing one round of DNA replication followed by two divisions. Homologous chromosomes are separated in the first division, and sister chromatids are separated in the second division. Both of these cell division cycles are used in the process of sexual reproduction at some point in their life cycle. Both are believed to be present in the last eukaryotic common ancestor.

Prokaryotes (bacteria and archaea) usually undergo a vegetative cell division known as binary fission, where their genetic material is segregated equally into two daughter cells, but there are alternative manners of division, such as budding, that have been observed. All cell divisions, regardless of organism, are preceded by a single round of DNA replication.

For simple unicellular microorganisms such as the amoeba, one cell division is equivalent to reproduction – an entire new organism is created. On a larger scale, mitotic cell division can create progeny from multicellular organisms, such as plants that grow from cuttings. Mitotic cell division enables sexually reproducing organisms to develop from the one-celled zygote, which itself is produced by fusion of two gametes, each having been produced by meiotic cell division. After growth from the zygote to the adult, cell division by mitosis allows for continual construction and repair of the organism. The human body experiences about 10 quadrillion cell divisions in a lifetime.

The primary concern of cell division is the maintenance of the original cell's genome. Before division can occur, the genomic information that is stored in chromosomes must be replicated, and the duplicated genome must be cleanly divided between progeny cells. A great deal of cellular infrastructure is involved in ensuring consistency of genomic information among generations.

In bacteria

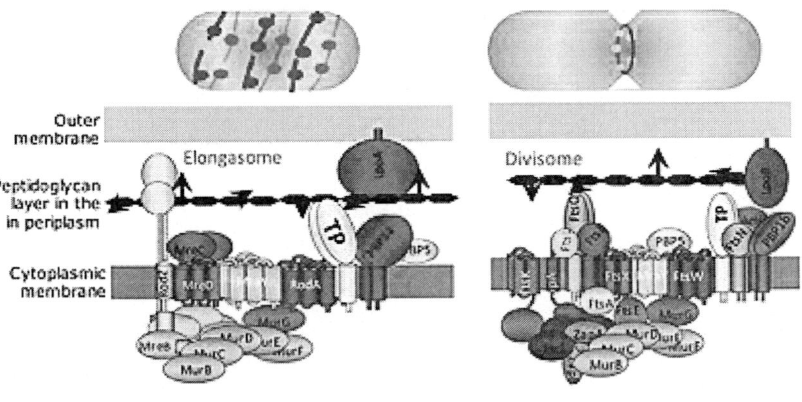

Divisome and elongasome complexes responsible for peptidoglycan synthesis during lateral cell-wall growth and division.

Bacterial cell division happens through binary fission or sometimes through budding. The divisome is a protein complex in bacteria that is responsible for cell division, constriction of inner and outer membranes during division, and remodeling of the peptidoglycan cell wall at the division site. A tubulin-like protein, FtsZ plays a critical role in formation of a contractile ring for the cell division.

Cell cycle (Interphase and mitosis)

A cell cycle is a series of events that takes place in a cell as it grows and divides. A cell spends most of its time in what is called interphase, and during this time it grows, replicates its chromosomes, and prepares for cell division. The cell then leaves interphase, undergoes mitosis, and completes its division. The resulting cells, known as daughter cells, each enter their own interphase and begin a new round of the cell cycle.

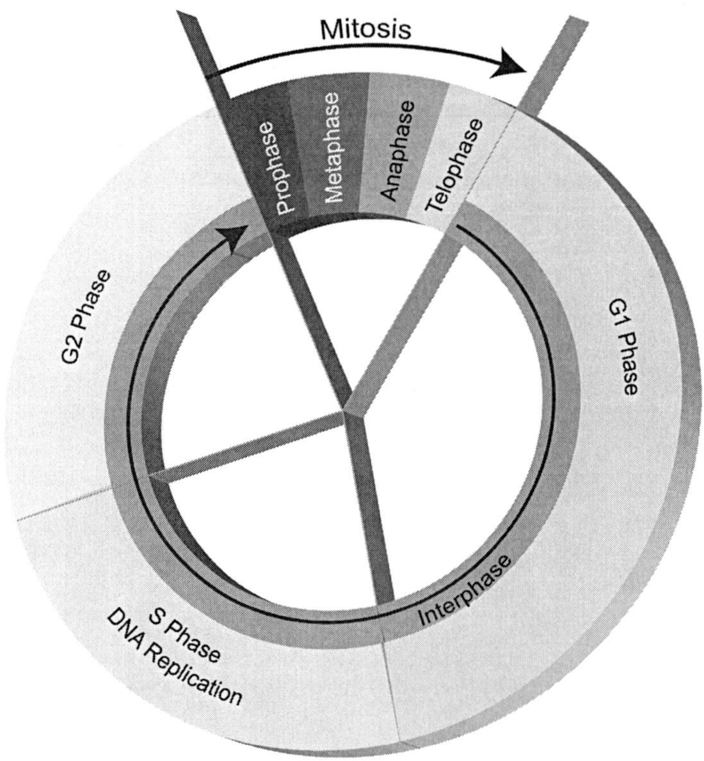

Cell cycle is the name we give the process through which cells replicate and make two new cells. Cell cycle has different stages called G1, S, G2, and M. G1 is the stage where the cell is preparing to divide. To do this, it then moves into the S phase where the cell copies all the DNA. So, S stands for DNA synthesis. After the DNA is copied and there's a complete extra set of all the genetic material, the cell moves into the G2 stage, where it organizes and condenses the genetic material, or starts to condense the genetic material, and prepares to divide. The next stage is M. M stands for mitosis. This is where the cell actually partitions the two copies of the genetic material into the two daughter cells. After M phase completes, cell division occurs and two cells are left, and the cell cycle can begin again.

Interphase

During interphase, the cell undergoes normal processes while also preparing for cell division. For a cell to move from interphase to the mitotic phase, many internal and external conditions must be met. The three stages of interphase are called G_1, S, and G_2.

G_1 Phase

The first stage of interphase is called the G_1 phase, or first gap, because little change is visible. However, during the G_1 stage, the cell is quite active at the biochemical level. The cell is accumulating the building blocks of chromosomal DNA and the associated proteins, as well as accumulating enough energy reserves to complete the task of replicating each chromosome in the nucleus.

S Phase

Throughout interphase, nuclear DNA remains in a semi-condensed chromatin configuration. In the S phase (synthesis phase), **DNA replication** results in the formation of two identical copies of each chromosome—sister chromatids—that are firmly attached at the centromere region. At this stage, each chromosome is made of two sister chromatids and is a duplicated chromosome. The centrosome is duplicated during the S phase. The two centrosomes will give rise to the mitotic spindle, the apparatus that orchestrates the movement of chromosomes during mitosis. The centrosome consists of a pair of rod-like centrioles at right angles to each other. Centrioles help organize cell division. Centrioles are not present in the centrosomes of many eukaryotic species, such as plants and most fungi.

G_2 Phase

In the G_2 phase, or second gap, the cell replenishes its energy stores and synthesizes the proteins necessary for chromosome manipulation. Some cell organelles are duplicated, and the cytoskeleton is dismantled to provide resources for the mitotic spindle. There may be additional cell growth during G_2. The final preparations for the mitotic phase must be completed before the cell is able to enter the first stage of mitosis.

The Mitotic Phase

To make two daughter cells, the contents of the nucleus and the cytoplasm must be divided. The mitotic phase is a multistep process during which the duplicated chromosomes are aligned, separated, and moved to opposite poles of the cell, and then the cell is divided into **two new identical daughter cells**. The first portion of the mitotic phase, mitosis, is composed of five stages, which accomplish nuclear division. The second portion of the mitotic phase, called cytokinesis, is the physical separation of the cytoplasmic components into two daughter cells.

Mitosis

Mitosis is divided into a series of phases—prophase, prometaphase, metaphase, anaphase, and telophase—that result in the division of the cell nucleus.

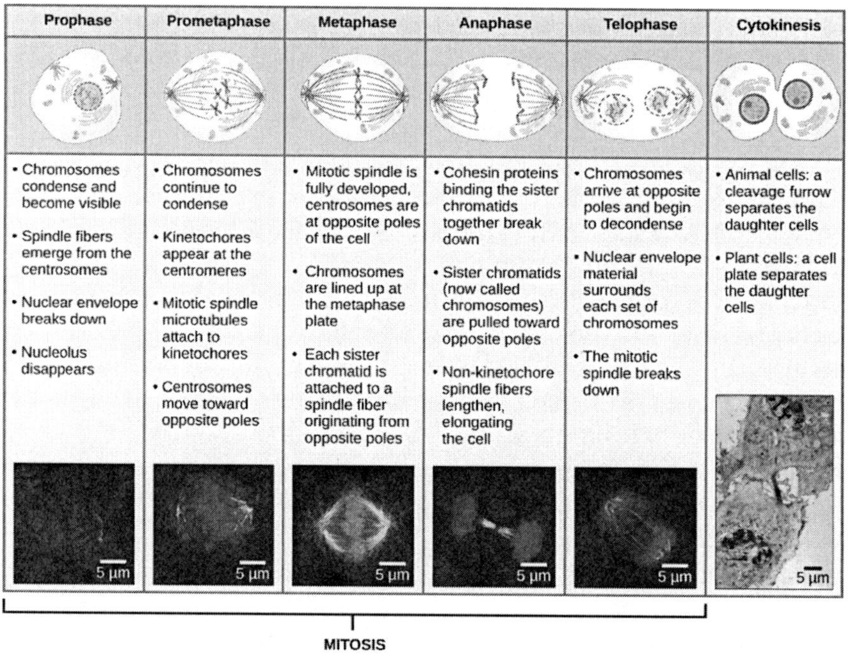

Figure : Animal cell mitosis is divided into five stages—prophase, prometaphase, metaphase, anaphase, and telophase—visualized here by light microscopy with fluorescence. Mitosis is usually accompanied by cytokinesis, shown here by a transmission electron microscope. (credit "diagrams": modification of work by Mariana Ruiz Villareal; credit "mitosis micrographs": modification of work by Roy van Heesbeen; credit "cytokinesis micrograph": modification of work by the Wadsworth

Center, NY State Department of Health; donated to the Wikimedia foundation; scale-bar data from Matt Russell)

Which of the following is the correct order of events in mitosis?

1. Sister chromatids line up at the metaphase plate. The kinetochore becomes attached to the mitotic spindle. The nucleus re-forms and the cell divides. The sister chromatids separate.
2. The kinetochore becomes attached to the mitotic spindle. The sister chromatids separate. Sister chromatids line up at the metaphase plate. The nucleus re-forms and the cell divides.
3. The kinetochore becomes attached to metaphase plate. Sister chromatids line up at the metaphase plate. The kinetochore breaks down and the sister chromatids separate. The nucleus re-forms and the cell divides.
4. The kinetochore becomes attached to the mitotic spindle. Sister chromatids line up at the metaphase plate. The kinetochore breaks apart and the sister chromatids separate. The nucleus re-forms and the cell divides.

During prophase, the "first phase," several events must occur to provide access to the chromosomes in the nucleus. The nuclear envelope starts to break into small vesicles, and the Golgi apparatus and endoplasmic reticulum fragment and disperse to the periphery of the cell. The nucleolus disappears. The centrosomes begin to move to opposite poles of the cell. The microtubules that form the basis of the mitotic spindle extend between the centrosomes, pushing them farther apart as the microtubule fibers lengthen. The sister chromatids begin to coil more tightly and become visible under a light microscope.

During prometaphase, many processes that were begun in prophase continue to advance and culminate in the formation of a connection between the chromosomes and cytoskeleton. The remnants of the nuclear envelope disappear. The mitotic spindle continues to develop as more microtubules assemble and stretch across the length of the former nuclear area. Chromosomes become more condensed and visually discrete. Each sister chromatid attaches to spindle microtubules at the centromere via a protein complex called the kinetochore.

During metaphase, all of the chromosomes are aligned in a plane called the metaphase plate, or the equatorial plane, midway between the two poles of the cell. The sister chromatids are still tightly attached to each other. At this time, the chromosomes are maximally condensed.

During anaphase, the sister chromatids at the equatorial plane are split apart at the centromere. Each chromatid, now called a chromosome, is pulled rapidly toward the centrosome to which its microtubule was attached. The cell becomes visibly elongated as the non-kinetochore microtubules slide against each other at the metaphase plate where they overlap.

During telophase, all of the events that set up the duplicated chromosomes for mitosis during the first three phases are reversed. The chromosomes reach the opposite poles and begin to decondense (unravel). The mitotic spindles are broken down into monomers that will be used to assemble cytoskeleton components for each daughter cell. Nuclear envelopes form around chromosomes.

Cytokinesis

Cytokinesis is the second part of the mitotic phase during which cell division is completed by the **physical separation of the cytoplasmic components** into two daughter cells. Although the stages of mitosis are similar for most eukaryotes, the process of cytokinesis is quite different for eukaryotes that have cell walls, such as plant cells.

In cells such as animal cells that lack cell walls, cytokinesis begins following the onset of anaphase. A contractile ring composed of actin filaments forms just inside the plasma membrane at the former metaphase plate. The actin filaments pull the equator of the cell inward, forming a fissure. This fissure, or "crack," is called the cleavage furrow. The furrow deepens as the actin ring contracts, and eventually the membrane and cell are cleaved in two.

In plant cells, a cleavage furrow is not possible because of the rigid cell walls surrounding the plasma membrane. A new cell wall must form between the daughter cells. During interphase, the Golgi apparatus accumulates enzymes, structural proteins, and glucose molecules prior to breaking up into vesicles and dispersing throughout the dividing cell. During telophase, these Golgi vesicles move on microtubules to collect at the metaphase plate. There, the vesicles fuse from the center toward the cell walls; this structure is called a cell plate. As more vesicles fuse, the cell plate enlarges until it merges with the cell wall at the periphery of the cell. Enzymes use the glucose that has accumulated between the membrane layers to build a new cell wall of cellulose. The Golgi membranes become the plasma membrane on either side of the new cell wall.

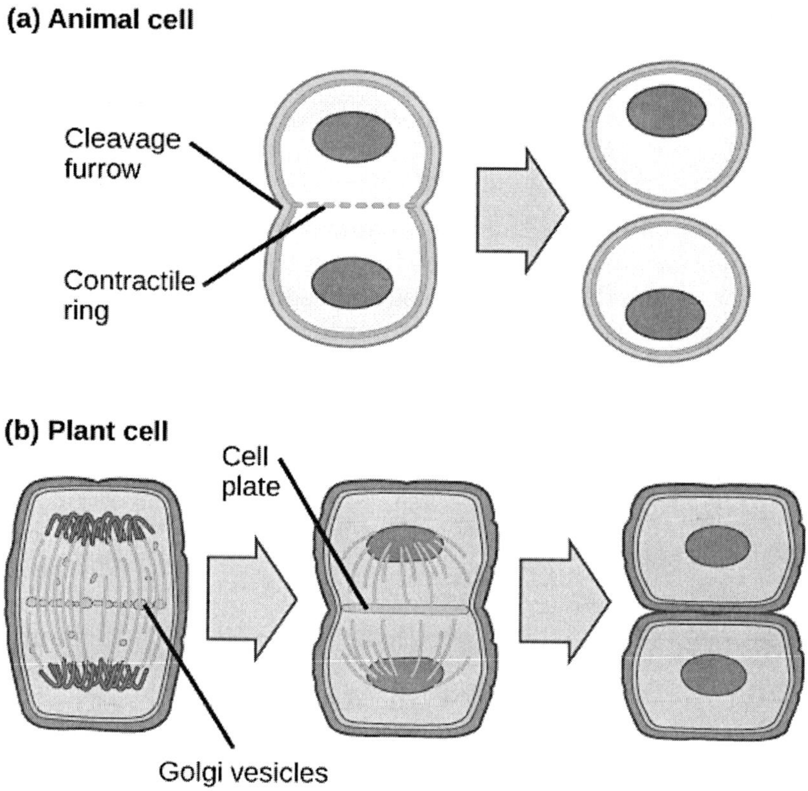

Figure : In part (a), a cleavage furrow forms at the former metaphase plate in the animal cell. The plasma membrane is drawn in by a ring of actin fibers contracting just inside the membrane. The cleavage furrow deepens until the cells are pinched in two. In part (b), Golgi vesicles coalesce at the former metaphase plate in a plant cell. The vesicles fuse and form the cell plate. The cell plate grows from the center toward the cell walls. New cell walls are made from the vesicle contents.

G_0 Phase

Not all cells adhere to the classic cell-cycle pattern in which a newly formed daughter cell immediately enters interphase, closely followed by the mitotic phase. Cells in the G_0 phase are **not actively preparing to divide**. The cell is in a quiescent (inactive) stage, having exited the cell cycle. Some cells enter G_0 temporarily until an external signal triggers the onset of G_1. Other cells that never or rarely divide, such as mature cardiac muscle and nerve cells, remain in G_0 permanently.

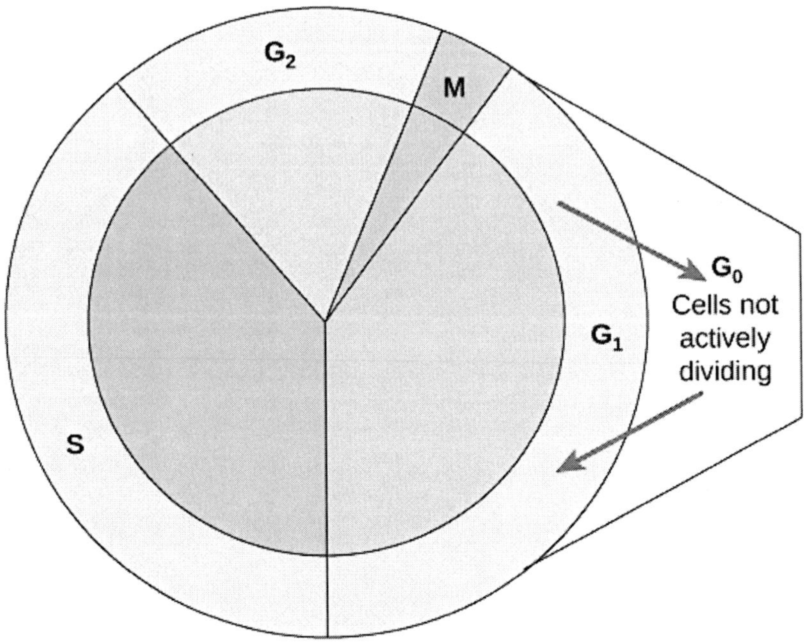

Figure : Cells that are not actively preparing to divide enter an alternate phase called G0. In some cases, this is a temporary condition until triggered to enter G1. In other cases, the cell will remain in G0 permanently.

Control of the Cell Cycle

The length of the cell cycle is highly variable even within the cells of an individual organism. In humans, the frequency of cell turnover ranges from a few hours in early embryonic development to an average of two to five days for epithelial cells, or to an entire human lifetime spent in G_0 by specialized cells such as cortical neurons or cardiac muscle cells. There is also variation in the time that a cell spends in each phase of the cell cycle. When fast-dividing mammalian cells are grown in culture (outside the body under optimal growing conditions), the length of the cycle is approximately 24 hours. In rapidly dividing human cells with a 24-hour cell cycle, the G_1 phase lasts approximately 11 hours. The timing of events in the cell cycle is controlled by mechanisms that are both internal and external to the cell.

Meiosis - significance of meiosis reproductive

Significance of Meiosis

Meiosis is the process in which the parent cell divides twice into four daughter cells containing half the original amount of genetic information, i.e., the daughter cells are haploid. The gametes are produced by meiosis.

Features of Meiosis

- It results in the formation of four daughter cells in each cycle of cell division.
- The daughter cells are identical to the mother cell in shape and size but different in chromosome number.
- The daughter cells are haploid.
- Recombination and segregation take place in meiosis.
- The process occurs in the reproductive organs and results in the formation of gametes.
- The process is divided into two types-Meiosis-I reduces the chromosome number to half and is known as reductional division. Meiosis-II is just like the mitotic division.

Significance

1. Meiosis is responsible for the formation of sex cells or gametes that are responsible for sexual reproduction.
2. It activates the genetic information for the development of sex cells and deactivates the sporophytic information.
3. It maintains the constant number of chromosomes by halving the same. This is important because the chromosome number doubles after fertilization.
4. In this process independent assortment of maternal and paternal chromosomes takes place. Thus the chromosomes and the traits controlled by them are reshuffled.
5. The genetic mutation occurs due to irregularities in cell division by meiosis. The mutations that are beneficial are carried on by natural selection.
6. Crossing over produces a new combination of traits and variations.

Explain the Significance of Meiosis in Sexual Reproduction

Species that reproduce sexually create offspring that inherit the genetic material of two parents. To prepare for sexual reproduction, an organism must create sex cells, or gametes, that have the ability to fuse with gametes from another individual during fertilization. Meiosis is the process that creates gametes and is therefore essential to sexual reproduction in eukaryotic organisms — that is, organisms such as humans with cells containing an organized nucleus.

Genetic Material

Chromosomes are the cellular structures that maintain and transmit genetic information. They are made of deoxyribonucleic acid (DNA) combined with proteins. The sequence of certain molecules, called nucleotide bases, within long strands of DNA codes for the sequence of amino acids in all of the proteins that an organism produces. The DNA is therefore responsible for all of the physical, and some of the behavioral, traits of an individual. These traits are a unique and random mix of those of the parents, determined by which particular sperm happens to fertilize the egg.

Ploidy

Cells in the human body contain 23 types of chromosomes. The human diploid number is 46, also expressed as 2N, which represents two copies of each chromosome, one copy from each parent. Gametes are special cells that have undergone ploidy reduction and contain only the haploid, or 1N, number of chromosomes. The gametes are the female egg and the male sperm that fuse together during fertilization, thereby restoring the 2N ploidy to the fertilized cell and initiating the events that result in embryo development and childbirth.

Meiosis I

Meiosis occurs over two cycles of cell division. In meiosis I, a cell's DNA replicates and forms chromosomes with two attached chromatids, each holding a full copy of the chromosome's genetic information. The cell nucleus dissolves and the 23 pairs of chromosomes line up along the center of the cell. Some pair members from each parent exchange portions of their DNA in a process called crossover that helps increase genetic diversity by creating non-identical chromatids. Following crossover, one double-chromatid member of each pair is pulled to one side of the cell while the other pair member is dragged to the opposite side. The spindle apparatus consisting of microtubules is responsible for pulling the chromosomes to each end of the cell.

Meiosis II

Meiosis I ends when the cell divides into two daughter cells, each having the haploid number of chromosomes, although each chromosome has two chromatids. After a resting phase, the cell enters meiosis II, in which the chromosomes once again align at the center of the cell. This time, the connections between the chromatids dissolve and the single-copy chromosomes are distributed to two daughter cells known as gametocytes. In this way, one cell results in four non-identical gametocytes that undergo further development to become sperm or eggs. Meiosis II in human females is delayed until fertilization.

Cycle stage of meiosis chiasma terminalization

The meiosis I comprises of following stages- Prophase, metaphase, anaphase, telophase. The prophase I comprises of leptotene, zygotene, pachytene, diplotene, diakinesis.

Terminalization of chiasma occurs during diplotene, after crossing over at pachytene, and terminalization completion takes place in diakinesis.

Meiosis is a specialized type of cell division. It involves a reduction in the number of chromosomes in daughter cells than the parent cells. This type of meiosis is evident in sexually reproducing organisms. The gametes which are mediators for sexual reproduction are made by this type of cell division. It consists of various stages and each stage is characterized by a special event occurring in it eventually causing cells to divide. The cell divides twice forming four daughter cells.

Synaptonemal complex & its significance in meiosis

Synaptonemal complex

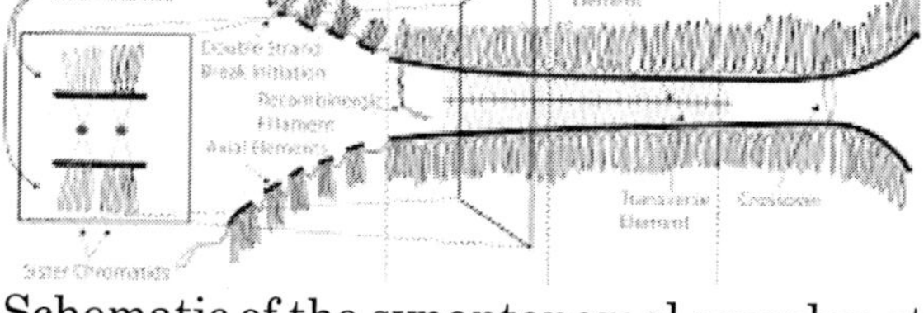

Schematic of the synaptonemal complex at different stages during Prophase I

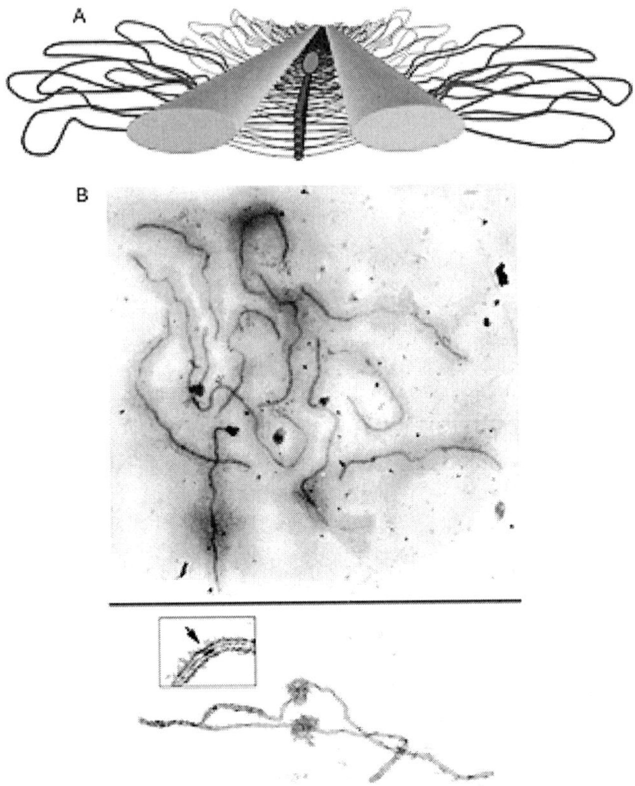

A Homologous chromosomes (light blue) align and synapse together via transverse filaments (black lines) and longitudinal filaments (dark blue). Recombination nodules (gray ellipsoids) on the central region may help in completing recombination. Chromatin (red loops) is attached to its sexual leg and toe, extending from both sister chromatids. **B** Top: Set of tomato SCs. Chromatin "sheaths" visible around each SC. Bottom: Two tomato SCs with the chromatin removed, allowing kinetochores ("ball-like" structures) at centromeres to be revealed.

The **synaptonemal complex** (**SC**) is a protein structure that forms between homologous chromosomes (two pairs of sister chromatids) during meiosis and is thought to mediate synapsis and recombination during meiosis I in eukaryotes. It is currently thought that the SC functions primarily as a scaffold to allow interacting chromatids to complete their crossover activities.

Composition

The synaptonemal complex is a tripartite structure consisting of two parallel lateral regions and a central element. This "tripartite structure" is seen during the pachytene stage of the first meiotic prophase, both in males and in females during gametogenesis. Previous to the pachytene stage, during leptonema, the lateral elements begin to form and they initiate and complete their pairing during the zygotene stage. After pachynema ends, the SC usually becomes disassembled and can no longer be identified.

In humans, three specific components of the synaptonemal complex have been characterized: SC protein-1 (SYCP1), SC protein-2 (SYCP2), and SC protein-3 (SYCP3). The SYCP1 gene is on chromosome 1p13; the SYCP2 gene is on chromosome 20q13.33; and the gene for SYCP3 is on chromosome 12q.

The synaptonemal complex was described by Montrose J. Moses in 1956 in primary spermatocytes of crayfish and by D. Fawcett in spermatocytes of pigeon, cat and man. As seen with the electron microscope, the synaptonemal complex is formed by two "lateral elements", mainly formed by SYCP3 and secondarily by SYCP2, a "central element" that contains at least two additional proteins and the amino terminal region of SYCP1, and a "central region" spanned between the two lateral elements, that contains the "transverse filaments" composed mainly by the protein SYCP1.

The SCs can be seen with the light microscope using silver staining or with immunofluorescence techniques that label the proteins SYCP3 or SYCP2.

Assembly and Disassembly

Formation of the SC usually reflects the pairing or "synapsis" of homologous chromosomes and may be used to probe the presence of pairing abnormalities in individuals carrying chromosomal abnormalities, either in number or in the chromosomal structure. The sex chromosomes in male mammals show only "partial synapsis" as they usually form only a short SC in the XY pair. The SC shows very little structural variability among eukaryotic organisms despite some significant protein differences. In many organisms the SC carries one or several "recombination nodules" associated to its central space. These nodules are thought to correspond to mature genetic recombination events or "crossovers". In male mice, gamma irradiation increases meiotic crossovers in SCs. This indicates that exogenously caused DNA damages are likely repaired by crossover recombination in SCs. The finding of an interaction between a SC structural component [synaptonemal central element protein 2 (SYCE2)] and recombinational repair protein RAD51 also suggests a role of the SC in DNA repair.

In cell development the synaptonemal complex disappears during the late prophase of meiosis I.It is formed during zygotene

Necessity in Eukaryotes

It is now evident that the synaptonemal complex is not required for genetic recombination in some organisms. For instance, in protozoan ciliates such as *Tetrahymena thermophila* and *Paramecium tetraurelia* genetic crossover does not appear to require synaptonemal complex formation. Research has shown that not only does the SC form after genetic recombination but mutant yeast cells unable to assemble a synaptonemal complex can still engage in the exchange of genetic information. However, in other organisms like the *C. elegans* nematode, formation of chiasmata require the formation of the synaptonemal complex.

Cytological basis of meiosis

Cytological basis of crossing over

Discussion in preceding sections of this section is based on the study of progeny in BC_1 generation or the testcross progeny. In such progenies, lack of independent assortment demonstrated linkage and presence of recombinants showed crossing over. It was concluded that crossing over should be the result of exchange of chromosome segments. However, since homologous chromosomes normally exchange reciprocal segments, resulting chromosomes will exhibit no morphological differences. Therefore, except the visible chiasmata, there is no other cytological observation which will substantiate that actual exchange of chromosome segments really takes place.

However, it is not necessary that chiasmata should be associated with exchange of chromosome segments. Therefore, to demonstrate that crossing over is associated with actual exchange of chromosome segments, special experiments were devised by G. Stern in *Drosophila* and by **H.S. Creighton** and **B. McCHntock** in corn. These experiments were reported in 1931 and had utilized chromosomes, whose morphology was altered due to chromosomal aberrations in order to make it identifiable from its homologue. Another experiment was conducted by M. Meselson and J. Weigle in lambda *(ë)* phage, demonstrating exchange of DNA segments.

Stern's experiment in Drosophila

Drosophila (fruitfly) stocks carrying translocations were utilised by **C. Stern** in order to produce a female *Drosophila* having a part of Y-chromosome attached to one of the two X-chromosomes. The **other** X-chromosome of this female fly was also marked (identifiable due to distinct morphology) and consisted of two approximately equal fragments, each carrying its centromere. These two X-chromosomes in the female fly could be distinguished not only from each other, but also from normal X-chromosome under the microscope.

In the above female fly, one of the two fragments of an X-chromosome carried mutant alleles for *carnation* eye (*car* is recessive showing light eye colour) and *barred* eye (*B* is dominant showing narrower eyes). The other X-chromosome, having a part of Y attached, carried normal alleles of these two genes, so that the female heterozygote for both these genes (*car B*/+ +) had barred eyes (but normal eye colour, since *car* is recessive to +). Such females were crossed with male flies having recessive alleles for both these genes (*car*, +). In such a situation, this makes a simple testcross. If no crossing over takes place between the two genes in question, two types of gametes i.e., *car B* and ++ will be produced from the female flies. Crossing over will give two additional types ot gametes i.e., *car* + and + *B*. Due to fertilization of two types of non-crossover and the other two types of crossover gametes by male gametes carrying X-chromosome (*car* +), four kinds of female flies will be produced. Another four kinds of male flies will be produced due to fertilization by Y carrying male gametes.

The flies which are classified as crossovers on the basis of phenotype i.e., carnation (with normal eye shape) and barred (with normal eye colour) were studied cytologically. It was found that carnation flies did not have any fragmented X-chromosome, but rather had normal X-chromosome. On the other hand, barred flies had a fragmented X-chromosome with a segment of Y-chromosome attached to one of the two fragments of X-chromosome. Such cytological observations suggested that genetic crossing over was accompanied with an actual exchange of chromosome segments.

Chapter 7

Linkage and Crossing Over

Introduction

In Eukaryotes, crossing over and genetic linkage are two different but related events. Crossing over is the phenomenon where the genetic material gets exchanged in the germline. Genetic linkage is the tendency of genes to stay together in a chromosome. Read on to explore how these two processes vary from each other and if they possess any similarities.

Genetic Linkage

Genetic Linkage means the tendency of the genes or DNA sequences on a chromosome to be inherited together during meiosis of sexual reproduction. Linked genes are the ones located on the same chromosome. For Example, genes on hair colour and eye colour. That is why some individuals inherit hair and eye colours together. Such as black hair with brown eyes or individuals with brown hair and blue eyes inherited from parents.

Crossing Over

The exchange of chromosomes between non-sister chromatids to form gametes is Crossing Over. Crossing Over shuffles the alleles on parental chromosomes, so that the gametes carry the combination of genes from mother and father. In simple language, it is the swapping of genetic material in the germline. During meiosis, or the formation of an egg cell and sperm cells, paired chromosomes get aligned so that similar DNA sequences from the paired chromosomes come against one another. This phenomenon accounts for genetic variation in sexually reproducing organisms and it is also essential for the normal segregation of chromosomes.

The Difference Between Linkage and Crossing Over

- Crossing Over is the process of separation of genes between homologous pairs into various gametes. Linkage is the tendency of inheriting genes together on the same chromosome.
- Linkage occurs when two genes are closer to each other on the same chromosome. On the other hand, Crossing Over takes place when two genes are located far apart on the same chromosome.
- Crossing Over may disrupt the gene groups made by Linkage.
- Unlike the Linkage, Crossing Over only occurs during the prophase of meiosis I and it produces recombinant alleles, which is not the case in the Linkage.

Coupling and repulsion Hypothesis

Bateson and Punnett in 1906, described a cross in sweat pea, where failure of gene pairs to assort independently was exhibited. Plants of a sweat pea variety having blue flower (BB) and long pollen (LL) were crossed with those of ano-ther variety having red flower (bb) and round pollen (II). F_1 individuals (BbLl) had blue flower and long pollen.

These were test crossed with plants having red flower and round pollen (bbll).

In this case, the character for blue colour of flower is dominant over red colour, and long pollen character is dominant over round pollen. In case of independent assortment, one should expect 1:1:1:1 ratio in test cross. Instead 7:1:1:7 ratio was actually obtained, indicating that there was a tendency of dominant alleles to remain together.

The case was similar for reces-sive alleles also. This deviation was explained in terms of gametic coupling by Bateson.

It was also observed that when two such dominant alleles or recessive alleles come from different parents, they tend to remain separate. This was termed as gametic repulsion. In Bateson's experiment, in repulsion phase, one parent would have blue flower and round pollen (BBII) and the other would have red flower and long pollen (bbLL).

The results of test cross in such a repulsion phase were similar to those obtained in coupling phase, giving 1:7:7:1 ratio instead of expected 1:1:1:1.

Bateson explained the lack of independent assortment by means of a hypothesis known as coupling and repulsion hypothesis. Later on, coupling and repulsion were discovered to be two different aspects of linkage.

Coupling and repulsion hypothesis

W. Bateson, in 1905, described a cross in sweet pea, where a deviation from independent assortment was exhibited. Plants of a sweet pea variety having **blue flowers *(B)*** and **long pollen** (L) were crossed with those of another variety having **red flowers *(b)*** and **round pollen** (*l*). F_1 individuals (*BbLl*) had blue flowers and long pollen. These were crossed with plants having red flowers and round pollen *(bbll)*. (In this case character for blue colour of flowers is dominam over red colour and long pollen character is dominant over round pollen).

Normally if independent assortment takes place, we should expect 1:1:1:1 ratio in a testcross. Instead, 7:1:1:7 ratio was actually obtained, indicating that there was a tendency in dominant alleles to remain together. Similar was the case with recessive alleles. This deviation was, therefore, explained as **gametic coupling by Bateson.** Similarly, it was observed that when two such dominant alleles or two recessive alleles come from different parents, they tend to remain separate. This was called **gametic repulsion.**

In Bateson's experiment in repulsion phase, one parent would have **blue flowers** and **round pollen** *(BBll)* and the other would have **red flowers** and **long pollen** *(bbLL)*. The results of a testcross in such a repulsion phase were similar to those obtained in coupling phase giving 1:7:7:1 ratio instead of expected 1:1:1:1. Bateson explained the lack of independent assortment in the above experiments by means of a hypothesis known as **coupling and repulsion hypothesis.** Although coupling and repulsion as explained above were later discovered to be the two aspects of the same phenomenon called **linkage,** the terms **coupling phase** and **repulsion phase** are still considered to be useful terms in scientific literature.

T.H. Morgan, while performing an experiment with *Drosophila,* in 1910, found that coupling or repulsion was not complete. He proposed that two genes are found in coupling phase or in repulsion phase, because these were present on same chromosome (coupling) or on two different, but homologous chromosomes (repulsion). Such genes are then called **linked** and the phenomenon is called linkage. He further suggested that the strength of linkage will, be determined by distance between two genes in question. The greater this distance, lower will be the linkage strength. The linkage is broken down due to the phenomenon of **crossing over** occurring during meiosis. Crossing over will be relatively more frequent, if distance between two genes is more than in a case where the distance between two genes is less. The phenomenon of crossing over involves exchange of chromosome segments

In order to make the phenomenon of linkage and crossing over easily understandable, let us take another hypothetical example, where genes *A* and *B* are involved, *a* and *b* being their recessive alleles respectively. A cross *AB/AB* x *ab/ab* would give rise to a F_1 dihybrid *AB/ab*. This dihybrid will then be crossed with double recessive parent *ab/ab* to get the testcross progeny. The formation of gametes from dihybrid *(AB/ab)* in presence and in absence of crossing over. The derivation of test cross progeny is also represented.

Depending upon distance between any two genes which is inversely proportional to strength of linkage, non-crossovers will vary from 50%-100% (100% non-crossovers is a state where no crossing over takes place as in male *Drosophila*. The crossovers will similarly vary from 0-50% and will never exceed 50%. The easiest way of finding out the proportion of non-crossovers and crossovers is to make a testcross *(AB/ab* x *ab/ab)*. In such a situation, since phenotypic ratio will depend upon ratio of different kinds of gametes coming from F_1 *(AB/ab)*, the relative proportion of non-crossovers and, crossovers can be easily determined from phenotypic ratio. This is illustrated in the next section using an example from maize.

Difference Between Coupling and Repulsion

The **key difference** between coupling and repulsion is that **coupling refers to the linkage of two dominant or two recessive alleles while repulsion refers to the linkage of dominant alleles with recessive alleles.**

Based on his experiments, Gregor Mendel stated three laws of inheritance: the law of dominance, the law of segregation and the law of independent assortment. Law of independent assortment states that genes for different traits segregate independently of each other during the gamete formation. However, some experiments show failure in the independent assortment of genes. During the test crosses, the expected ratios are not observed. This is due to genetic linkage. Coupling and repulsion are two different aspects of linkage. Coupling refers to the linkage of two dominant or two recessive alleles while repulsion refers to the linkage of dominant alleles with recessive alleles.

What is Coupling?

Coupling is the linkage of two dominant alleles of two genes in one chromosome and two recessive alleles of two genes in the other homologous chromosome. Here, dominant alleles of the genes are present in one chromosome, while their recessive alleles are present in the other chromosome. These linked genes show *cis* arrangement. It can be illustrated as AB/ab.

Figure : Coupling Conformation

One chromosome carries AB and the other carries ab. This physical coupling between dominant alleles and between recessive alleles prevents their independent assortment during the gamete formation. Dominant alleles tend to remain together. Similarly, recessive alleles also tend to remain together during the gamete formation.

What is Repulsion?

Repulsion is another aspect of linkage which is different from coupling. In repulsion, dominant alleles or recessive alleles come from different parents, and they tend to remain separate. Here, one parental chromosome carries one dominant and one recessive allele while the other chromosome carries the other two alleles (dominant and recessive alleles). It can be illustrated as Ab/aB.

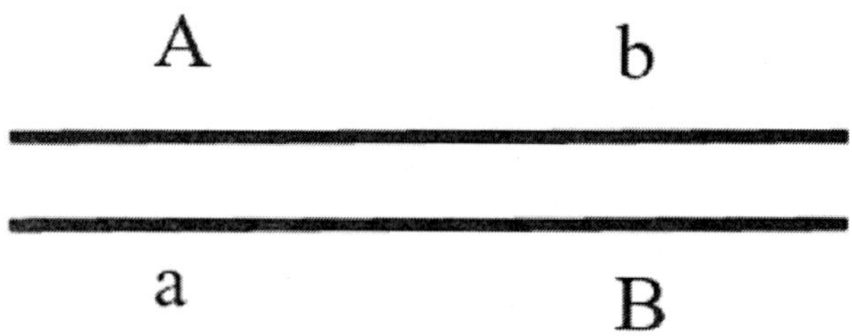

Figure : Repulsion Conformation

One dominant allele is linked with the recessive allele of the second gene. This type of gene arrangement is called *trans* arrangement.

What are the Similarities Between Coupling and Repulsion?

- Coupling and Repulsion are two aspects of linkage.
- They behave against Mendel's law of independent assortment.

What is the Difference Between Coupling and Repulsion?

In coupling, there is a tendency in dominant alleles to remain together; there is a tendency in recessive alleles to remain together. On the other hand, in repulsion, two such dominant alleles or two recessive alleles come from different parents, and they tend to remain separate. So, this is the key difference between coupling and repulsion. The observed ratio in coupling is 7:1:1:7 while the ratio of repulsion is 1:7:7:1. Moreover, coupling is a type of cis arrangement, while repulsion is a type of trans arrangement.

A Test-cross in maize crossing over

A testcross in maize

An example like the one for sweet pea outlined in the previous section is used in order to work out recombination frequencies. **C.B. Hutchinson** in 1922 made use of two characters known to be located on chromosome 9 in maize. These two characters were **coloured aleurone *(C)*** vs. **colourless aleurone *(c)*** and **full endosperm *(Sh)*** vs. **shrunken endosperm** *(sh)* in seed of maize crop. Seed characters have a special utility in inheritance studies, since in the next generation no crop needs to be raised to study the segregation; the characters are expressed at seed stage itself.

Hutchinson performed the experiment in two sets, once having two seed characters in coupling phase and the other having these characters in repulsion phase. The actual results obtained. The recombination frequencies can be worked out as per cent total population and would be 301/8,368 = 3.6% in coupling phase and would be 1310/44,595 = 2.94% in repulsion phase. It is obvious that strong linkage force exists between aleurone colour character and endosperm shape character.

Table: Results obtained in maize by Hutchinson in two testcrosses : CS/cs × cs/cs (coupling phase); Cs/cS × cs/cs (repulsion phase)

	Phenotypes and genotypes				
	Coloured full (*CS/cs*)	**Colour-less shrunken** (*cs/cs*)	**Coloured shrunken** (*Cs/cs*)	**Colour-less full** (*cS/cs*)	**Total**
(i) Coupling phase	4,032	4,035	149	152	8,368
(ii) Repulsion phase	638	672	21,379	21,906	44,595

Crossing over and linkage Maps

Multiple Point Gene Mapping

Gene mappers are motivated to map all of the tens of thousands of genes found on the chromosomes of plant or animals. Analyzing data from crosses to determine map distances for two genes at a time makes the process time consuming and tedious. Therefore geneticists will often attempt to map as many genes as possible from a single set of progeny. We will go through the simplest multiple point mapping example, a three point testcross, to demonstrate the process.

We will use our three corn seed trait loci again and use data from one cross to map these three loci. The first step would be to obtain a trihybrid individual that is heterozygous at all three loci and then perform a testcross with this trihybrid.

Parents: CCssWW (CsW/CsW) X ccSSww (cSw/cSw)

Trihybrid: CcSsWs (CsW/cSw) X ccssww (csw/csw)

Test cross offspring

Seed trait	Gamete from trihybrid	Number
Red, shrunken, normal	CsW	2777
White, plump, waxy	cSw	2708
Red, plump, waxy	CSw	116
White, shrunken, normal	csW	123
Red, shrunken, waxy	Csw	643
White, plump, normal	cSW	626
Red, plump, normal	CSW	4
White, shrunken, waxy	csw	3
Total number of progeny:		**7000**

We observe eight phenotypes of seeds in the testcross progeny because the trihybrid can make eight kinds of gametes. As we knew, these three genes are linked and so the uneven ratio of phenotypes reflects the combinations of two parental and six recombinant gametes. The parental gametes are still the most frequently produced type and the six recombinant gametes are made when crossing over occurs between the loci. We can systematically account for these crossovers if we follow the following four steps. First we will explain the steps and then we will show how to use them to map the three genes.

Step 1: Identify the parental gametes.

These are CsW and cSw, that combination which came from the parent generation and the combination made by the trihybrid at the highest frequency.

Step 2: Classify the recombinants.

If we observe the gene combination in each of the six recombinant gametes, we can ask ourselves if the gamete has a new combination for each pair of genes. For example CSw has a new combination for the CS and the Cw compared to the parental gametes but the same combination for Sw as the cSw parental gamete. Therefore, these 116 gametes represent crossovers between C,c and S,s as well as C,c and W,s. This information is recorded.

Step 3: Determine recombinant gamete frequency.

Once all six recombinant gametes have been classified, the total number of crossovers between the three loci can be added up and crossover percentage determined between each pair of loci. This number will reflect gene distance but one more step is needed to complete the process.

Step 4: Add in the double crossover gametes.

Two of the six recombinant gametes were made as a result of double crossovers between the two loci that are furthest apart. These crossovers have been added to the map distances between the middle locus and the two outside loci. Now we need to add these double crossovers to the outside loci distance.

Applying these fours steps to our three point test cross data would work as follows:

Step1:	CsW and cSw
Step 2:	C,c to S,s - 116 (CSw) + 123 (csW) + 4 (CSW) + 3 (csw) = 246
	C,c to W,w - 116 (CSw) + 123 (csW) + 643 (Csw) + 626 (cSW) = 1508
	S,s to W,w - 643 (Csw) + 626 (cSW) + 4 (CSW) + 3 (csw) = 1276
Step 3:	246 / 7000 = 3.5% recombinant gametes C,c to S,s 1508 / 7000 = 21.5% recombinant gametes C,c to W,w 1276 / 7000 = 18.2% recombinant gametes S,s to W,w
Step 4:	From the information in step 3 we can see that the C,c and W,w loci are the farthest apart so the S,s locus must be between them. We can also see that we have underestimated the distance between the outside loci (3.5 C,c to S,s + 18.2 S,s to W,w = 21.7 map units not 21.5). While this difference is small, we can rectify this and double check our work by adding in the double crossovers. The CSW and csw gametes are made very rarely. That is because is takes two crossovers in the trihybrid's chromosome to make them, not just one.

When two crossovers occur between the C,c and W,w loci the parental combinations of CW and cw are restored. Therefore we did not count the CSW and csw gametes as representing crossovers when they actually represent two crossovers each. Thus we should add up the double crossovers (3 + 4 = 7), multiply times two and add these 14 crossovers to the C,c to W,w distance (1508 + 14 = 1522 / 7000 = 21.7 map units). Now we have double checked our map distances and have our three point map complete.

It can be intimidating to see all the data generated from a three point test cross. This information, however, can be used systematically to save the geneticists time and map three genes in one experiment. The three point data also provides a more accurate measure of map distances compared to two point data when genes are farther apart on a chromosome. This is because a third gene in between the more distant loci can account for double crossovers that would not be detectable in a two point analysis. For this reason, map distances tend to be underestimated when genes are further apart as we saw to a modest extent in this example (21.5 two-point vs. 21.7 three-point).

How would a geneticist work with a four point test cross? There would be sixteen phenotypes in the progeny as a result of the hybrid parent making two parental gametes and fourteen recombinant gametes. These recombinant gametes would be a result of single, double and even triple crossovers that occurred in prophase I. While the data would be tedious to work with, one data set could be analyzed to reveal the map of four genes. It is not surprising that geneticists now have written computer programs which will perform these types of calculations, save time and reduce the chance of calculator error. Geneticists mapping genes in economically important plants and animals will also generate mapping populations that are a result of crossing parents that have different alleles at hundreds or thousands of loci. Even though the development of computers and programs has become a part of modern gene mapping the process of indirectly measuring crossover frequency by observing the inheritance of trait combinations is the basis of generating these gene maps.

Deriving Linkage Distance and Gene Order From Three-Point Crosses

By adding a third gene, we now have several different types of crossing over products that can be obtained. The following figure shows the different recombinant products that are possible.

Parental and Recombinant Gametes Derived from Three-Point Crosses

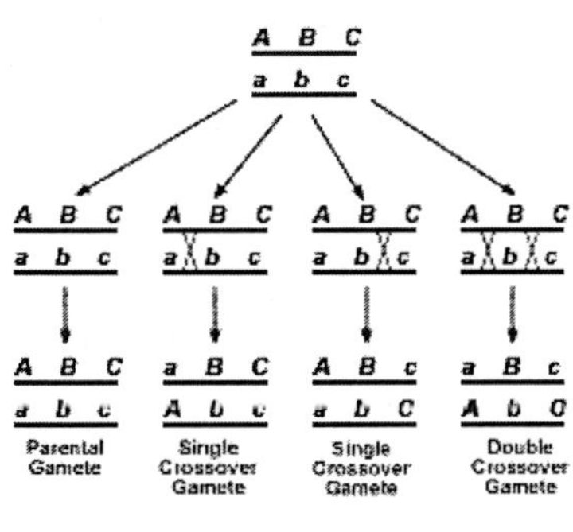

Now if we were to perform a testcross with F_1, we would expect a 1:1:1:1:1:1:1:1 ratio. As with the two-point analyzes described above, deviation from this expected ratio indicates that linkage is occurring. The best way to become familiar with the analysis of three-point test cross data is to go through an example. We will use the arbitrary example of genes *A*, *B*, and *C*. We first make a cross between individuals that are *AABBCC* and *aabbcc*. Next the F_1 is testcrossed to an individual that is *aabbcc*. We will use the following data to determine the gene order and linkage distances. As with the two-point data, we will consider the F_1 gamete composition.

Genotype	**Observed**	**Type of Gamete**
ABC	390	Parental
abc	374	Parental
AbC	27	Single-crossover between genes *C* and *B*
aBc	30	Single-crossover between genes *C* and *B*
ABc	5	Double-crossover
abC	8	Double-crossover
Abc	81	Single-crossover between genes *A* and *C*
aBC	85	Single-crossover between genes *A* and *C*
Total	**1000**	

The best way to solve these problems is to develop a systematic approach. First, determine which of the the genotypes are the parental gentoypes. The genotypes found most frequently are the parental genotypes.

Next we need to determine the order of the genes. Once we have determined the parental genotypes, we use that information along with the information obtained from the double-crossover. The double-crossover gametes are always in the lowest frequency. From the table the ABc and abC genotypes are in the lowest frequency. The next important point is that a double-crossover event moves the middle allele from one sister chromatid to the other. This effectively places the non-parental allele of the middle gene onto a chromosome with the parental alleles of the two flanking genes. We can see from the table that the *C* gene must be in the middle because the recessive *c* allele is now on the same chromosome as the *A* and *B* alleles, and the dominant *C* allele is on the same chromosome as the recessive *a* and *b* alleles.

Now that we know the gene order is *ACB*, we can go about determining the linkage distances between *A* and *C*, and *C* and *B*. The linkage distance is calculated by dividing the total number of recombinant gametes into the total number of gametes. This is the same approach we used with the two-point analyses that we performed earlier. What is different is that we must now also consider the double-crossover events. For these calculations we include those double-crossovers in the calculations of both interval distances.

So the distance between genes *A* and *C* is 17.9 cM [100*((81+85+5+8)/1000)], and the distance between *C* and *B* is 7.0 cM [100*((27+30+5+8)/1000)].

Now let's try a problem from *Drosophila*, by applying the principles we used in the above example. The following table gives the results we will analyze.

Genotype	Observed	Type of Gamete
$v\ cv^+\ ct^+$	580	Parental
$v^+ cv\ ct$	592	Parental
$v\ cv\ ct^+$	45	Single-crossover between genes *ct* and *cv*
$v^+ cv^+\ ct$	40	Single-crossover between genes *ct* and *cv*
$v\ cv\ ct$	89	Single-crossover between genes *v* and *ct*
$v^+ cv^+\ ct^+$	94	Single-crossover between genes *v* and *ct*
$v\ cv^+\ ct$	3	Double-crossover
$v^+ cv\ ct^+$	5	Double-crossover
Total	**1448**	

Step 1: Determine the parental genotypes.

The most abundant genotypes are the partenal types. These genotypes are $v\ cv^+ ct^+$ and $v^+ cv\ ct$. What is different from our first three-point cross is that one parent did not contain all of the dominant alleles and the other all of the recessive alleles.

Step 2: Determine the gene order

To determine the gene order, we need the parental genotypes as well as the double crossover geneotypes As we mentioned above, the least frequent genotypes are the double-crossover geneotypes. These geneotypes are $v\ cv^{+}\ ct$ and $v^{+}\ cv\ ct^{+}$. From this information we can determine the order by asking the question: In the double-crossover genotypes, which parental allele is not associated with the two parental alleles it was associated with in the original parental cross. From the first double crossover, $v\ cv^{+} ct$, the *ct* allele is associated with the *v* and cv^{+} alleles, two alleles it was not associated with in the original cross. Therefore, *ct* is in the middle, and the gene order is *v ct cv*.

Step 3: Determing the linkage distances.

- *v* - *ct* distance caluculation. This distance is derived as follows: 100*((89+94+3+5)/1448) = 13.2 cM
- *ct* - *cv* distance calculation. This distance is derived as follows: 100*((45+40+3+5)/1448) = 6.4 cM

Step 4. Draw the map.

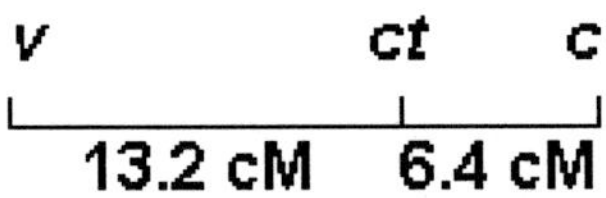

Three-point crosses also allows one to measure interference (I) among crossover events within a given region of a chromosome. Specifically, the amount of double crossover gives an indication if interference occurs. The concept is that given specific recombination rates in two adjacent chromosomal intervals, the rate of double-crossovers in this region should be equal to the product of the single crossovers. In the *v ct cv* example described above, the recombination frequency was 0.132 between genes *v* and *ct*, and the recombination frequency between *ct* and*cv* was 0.064. Therefore, we would expect 0.84% [100*(0.132 x 0.64)] double recombinants. With a sample size of 1448, this would amount to 12 double recombinants. We actually only detected 8.

To measure interference, we first calculate the coefficient of coincidence (c.o.c.) which is the ratio of observed to expected double recombinants. Interference is then calculated as 1 - c.o.c. The formula is as follows:

$$I = 1 - \text{c.o.c.} = 1 - \left[\frac{\text{Observed \# of double recombinants}}{\text{Expected \# of double recombinants}}\right]$$

For the *v ct cv*data, the interference value is 33% [100*(8/12)].

Most often, interference values fall between 0 and 1. Values less than one indicate that interference is occurring in this region of the chromosome.

Chapter 8

Genetics of Sex Determination

Introduction

Sex determination results in the development of individuals with characteristics that allow them to be identified as males, females, or in some cases, hermaphrodites. In certain species, like the solid nematode *C. elegans*, differences in sexual characteristics can be very small; in fact, in *C. elegans*, the only distinguishing sexual characteristic is the presence of a testis versus an ovotestis. In other species, the phenotypic differences between the sexes can be quite significant. Consider, for example, the remarkable plumage and display of a tom turkey versus the rather plain features of a female turkey. Female and male mammals are also readily distinguished by many differences in their internal and external phenotypes, behavior, and metabolism.

The first major breakthrough in understanding sex determination was the discovery of sex chromosomes in the early 1900s. From meticulous analyses of male and female insect chromosomes, scientists discovered that, although most chromosomes were present in equal numbers in both males and females, there were one or two additional chromosomes that were unequally represented in the two sexes. Analyses of additional species over the years has revealed that chromosomal differences are primarily responsible for sex determination in most animals.

Chromosomal Mechanisms of Sex Determination in Insects

Insects are the most diverse class of organisms on the planet, so it is not too surprising that they show considerable diversity in their mechanisms of sex determination (Saccone *et al.*, 2002). However, like most other animals, the majority of insects have dimorphic sex chromosomes that can be distinguished cytologically.

Animals with two different sex chromosomes are of heterogametic sex, and they are thus able to produce two types of gametes. Conversely, members of the homogametic sex can only produce one type of gamete. In humans (as well as many other animals), males generally have one X and one Y chromosome, while females have two X chromosomes. This system is reversed in butterflies and moths (order Lepidoptera): females are the heterogametic sex, while males are homogametic. The sex chromosomes in Lepidoptera are designated W and Z. W chromosome is usually associated with the development of female characteristics. When the W chromosome is absent, ZZ develops into males and ZO develop into females. Having a W chromosome to develop as a female isn't even a necessity for some species. A moth known as *Talaeporia tubulosa* uses the ambient temperature to control sex determination in the absence of W chromosome. When temperatures are warm, the Z chromosome is found on the inner spindle and more female eggs are produced, whereas in colder conditions, the Z chromosome moves to the outer pole, resulting in greater production of males (Traut *et al.*, 2007). This system corresponds to the adaptive advantage of favoring the production of female offspring when conditions are good (warm) and resources for their subsequent reproduction are more likely to be available.

Some grasshoppers also use a single-chromosome (XX/XO) sex determination system; here, males have only one sex chromosome, so they are considered to be XO. Thus, males are the heterogametic sex, because they produce two different kinds of gametes.

The system of chromosomal sex determination is even further reduced in certain genera of mosquitoes, in which the two sexes are chromosomally indistinguishable. Sex in this homogametic group is thought to be determined by a dominant male-determining factor.

Sex Determination in Drosophila

The sex chromosomes of the fruit fly Drosophila melanogaster have played a particularly important role in our understanding of heredity. Therefore, it may come as a surprise that fruit flies use a relatively rare mechanism to determine sex. In fact, in *Drosophila*, sex is primarily determined by the X:A ratio, or the ratio of the number of X chromosomes to the number of sets of autosomes (Cline & Meyer, 1996). The balance between female-determining factors encoded on the X chromosome and male-determining factors encoded on the autosomes determines which sex-specific pattern of transcription will be initiated. Thus, XX, XXY, and XXYY flies are females, while XY and XO flies are males. Flies are unable to survive with more than two copies of an X chromosome because of the mechanism that they use for dosage compensation. (Dosage compensation refers to the processes by which animals equalize the amount of gene products generated from X-linked genes in males and females. Unlike in mammals, all of the *Drosophila* X chromosomes remain active, and flies adjust the levels of X-linked gene products by doubling expression from the X chromosome in males. An extra copy of the X chromosome, which contains close to one-third of fly genes, creates an aneuploid condition that greatly disrupts the equilibrium in cells.)

Drosophila sex determination also differs from mammalian sex determination in several other ways. First, sex determination begins immediately at fertilization, and there is no indifferent period. Furthermore, hormones are not responsible for sex-specific traits; instead, each cell in the embryo senses the X:A ratio, triggering either the female- or male-specific pattern of transcription. Microarray experiments indicate that the sex-specific differences in gene expression are quite extensive. In fact, roughly 30% of *Drosophila* genes were found to show sex-specific biases in expression (Parisi *et al.*, 2004).

Sex Determination in Mammals

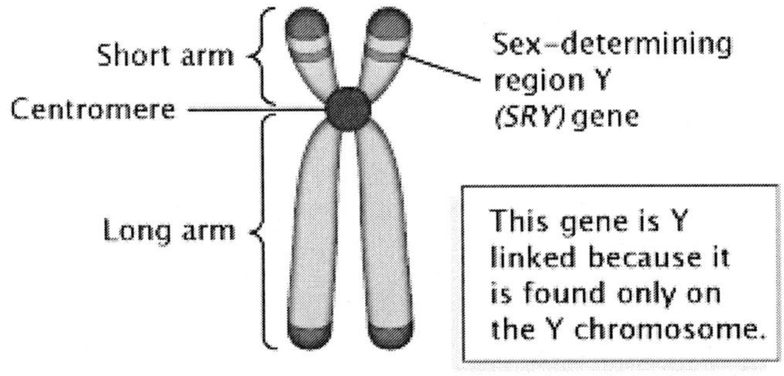

Figure : Location of the sex-determining region Y (SRY) gene on the Y-chromosome.

The SRY gene (blue band) on the male Y chromosome regulates sex determination in mammals.

In placental mammals, the presence of a Y chromosome determines sex. Normally, cells from females contain two X chromosomes, and cells from males contain an X and a Y chromosome. Occasionally, individuals are born with sex chromosome aneuploidies, and the sex of these individuals is always determined by the absence or presence of a Y chromosome. Thus, individuals with 47,XXY and 47,XYY karyotypes are males, while individuals with 45,X and 47,XXX karyotypes are females. Humans are able to tolerate supernumerary numbers of sex chromosomes because of X inactivation and the fact that the human Y chromosome is quite gene-poor.

Although the role of the Y chromosome in mammalian sex determination has been known since the early twentieth century, it was not until 1959 that scientists were able to identify the region of the Y chromosome that controlled this process (McLaren, 1991). Later, researcher David C. Page analyzed the chromosomes of sex-reversed XX men, rare individuals who look like men but have two X chromosomes instead of one X chromosome and one Y chromosome. Using DNA hybridization with probes corresponding to different regions of the Y chromosome, Page discovered that sex-reversed males carried genes from a 140-kilobase region on the short arm of the Y chromosome. Presumably, this region had been transferred to the X chromosome during a translocation (Page et al., 1985). Subsequent experiments narrowed down this region (McLaren, 1991) and found that one gene, the sex-determining region of the Y, or SRY, was the master regulator of sex determination. The presence of just this region from the Y chromosome is thus sufficient to cause male development (Koopman et al., 1991).

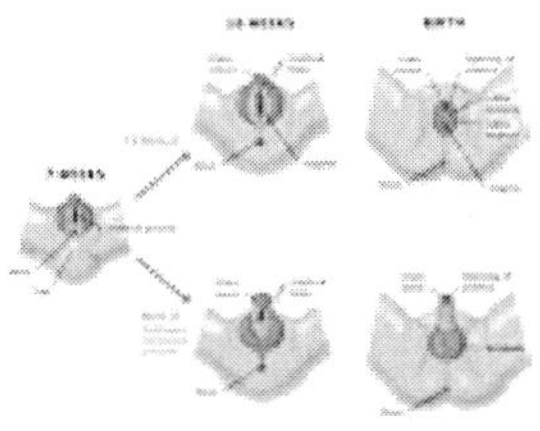

In human embryos, the SRY gene encodes a unique transcription factor that activates a testis-forming pathway at about week seven of development. Before this time, the embryonic gonad is "indifferent," meaning that it is capable of developing into either a testis or an ovary. Likewise, the early embryo has two systems of ducts, Wolffian and Müllerian ducts, which are capable of developing into the male and female reproductive tracts, respectively. Once the SRY gene product stimulates the indifferent gonad to develop into a testis, the testis begins producing two hormones, testosterone and anti-Müllerian hormone, or AMH. Testosterone and one of its derivatives, dihydrotestosterone, induce formation of other organs in the male reproductive system, while AMH causes the degeneration of the Müllerian duct. In females, who do not contain the SRY protein, the ovary-forming pathway is activated by a different set of proteins. The fully developed ovary then produces estrogen, which triggers development of the uterus, oviducts, and cervix from the Müllerian duct.

Sex Determination in Birds: Z and W Chromosomes

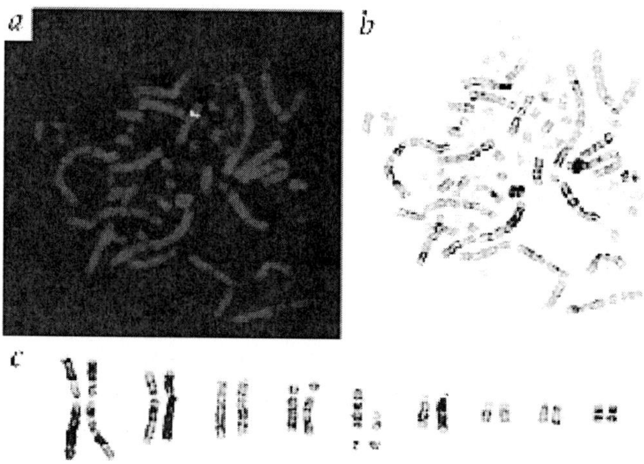

Figure : FISH mapping of DMRT1 on chicken Z chromosome.

a. Hybridization of DMRT1 cosmid to a female chicken metaphase spread. b. DAPI banding of the same metaphase spread, converted into G-like bands. c. G-banded karyotype of chicken macrosomes hybridized with DMRT1 (green spot on Z chromosome).

In birds, sex is determined by chromosomes known as the Z and W, and females are the heterogametic sex. Early on, it was apparent that there were notable differences in the mechanisms used for sex determination in birds and mammals. Working with chickens, scientists were unable to find a counterpart of the *SRY* gene required for mammalian testis determination, so they searched for homologues of other genes that were required for testis formation in mammals. These investigations led to the discovery of the *DMRT1* gene on the chicken Z chromosome (Nanda *et al.*, 1999).

In mammals, the *DMRT1* gene product is a critical member of the testis-forming pathway initiated by *SRY*. Two copies of the *DMRT1* gene are necessary for testis development, even in the presence of *SRY*. In mammals, both sexes have two copies of *DMRT1*, because it is located on an autosome (specifically, chromosome 9). In chickens, by contrast, only males possess two copies of *DMRT1*, as it is located on the Z chromosome. The results of an *in situ* hybridization experiment in which a *DRMT1* probe has been hybridized to the chromosomes of a female chicken. The appearance of the chromosomes under a microscope, and that the *DRMT1* probe hybridizes to a single region on the female's Z chromosome. (These are metaphase chromosomes, so hybridization is detected on the two sister chromatids.) Selected chromosomes in a partial karyotype. One can readily see that the Z chromosome is significantly larger than the W chromosome.

The actual trigger for activating the testis-forming pathway in chickens remains unknown. Birds also differ significantly from mammals in that two unique genes on the W chromosome, *FET1* and *ASW*, are necessary for female development. The function of *FET1* is unknown, but it is expressed in the gonads leading up to the time of sexual differentiation (Smith & Sinclair, 2004). Like mammals, chickens also have an indifferent gonad until around day four of development, after which an ovary or a testis starts to develop. Hormones then orchestrate the development of other sex-specific characteristics. In contrast with mammals, however, estrogen is required earlier in sex determination in chickens, and is, in fact, necessary for formation of the ovary. In fact, genetically male chickens can be converted to females if eggs are injected with estrogen at the sensitive stage of development.

The ZZ-ZW mechanism of sex determination is not restricted to birds. Within the vertebrates, a similar system of sex determination has been identified in reptiles, as well as in some fishes and amphibians.

Sex Determination and Environment

As previously mentioned in the discussion of *Talaeporia tubulosa*, environmental factors can sometimes play an important role in sex determination. Insects are not a special case—among the vertebrates, temperature also has a strong influence on sex determination in certain groups of reptiles. For example, in crocodilian reptiles and most turtles, sex is determined by egg incubation temperature. There are several variations on this theme. In the alligator snapping turtle, *Macroclemys temminckii*, incubation of eggs below 22°C or above 28°C gives rise to females, while incubation at intermediate temperatures produces predominantly males. American alligators show a similar biphasic dependence on temperature, but the curve is shifted to higher temperatures. In the European pond turtle, *Trachemys scripta*, incubation temperatures above 30°C produce all females, whereas incubation temperatures below 25°C produce all males. At 28.5°C, equal numbers of males and females are produced.

The temperature-dependent component of the sex-determination pathway has been studied in great detail in the European turtle, *Emys orbicularis*. In *E. orbicularis*, the critical temperature-dependent component appears to be synthesis of the enzyme aromatase, which converts androgens, such as testosterone, into estrogens. At higher temperatures, increased aromatase activity produces more estrogens, which biases the sex ratio toward more females.

As one compares the various mechanisms for sex determination among species, it is clear that evolution has produced numerous solutions for generating different sexes. Sexual reproduction has tremendous adaptive value to a species, because it introduces new genetic variability into a population in each new generation. Chromosomes play determinative roles in most species, but even so, environmental factors introduce additional wrinkles into the developmental process.

Chromosomal genotype sex Determination in plants and animals

Determination Of Sex

There are two different types of sexes, that participate in sexual reproduction. So it is natural to find it confounding as from which sex the baby inherits, that results in the sex of the child. There are several other procedures followed to determine the sex of a newborn baby. Based on the environmental signals the sex of a baby can be determined. In a few animal species, temperature plays a major role in sex determination. In other animals, like snails, it is possible to change sex as they are not genetically processed. In human beings, the sex of an individual is genetically determined. In other words, the genes which are inherited from their parents decide the sex of the child.

SEX DETERMINATION IN FLOWERING PLANTS

The hypothesis that angiosperms originated from a hermaphroditic ancestor is supported by the prevalence of hermaphrodites among about 72% of all species (**Takhtajan, 1969**). Diverse mechanisms that evolved to promote outcrossing include temporal separation of the maturation of male or female organs, gametophytic or sporophytic self-incompatibility, structural avoidance, monoecy, and ultimately dioecy. The first step toward sex chromosome evolution is the occurrence of male sterile or female sterile mutations leading to the development of unisexual flowers. Such mutations occur frequently and repeatedly in plant species, as demonstrated by the presence of unisexual species in 75% of angiosperm families and numerous male sterile mutants in domesticated crop plants (Yampolsky and Yampolsky, 1922; Renner and Ricklefs, 1995). Also, stamen and carpel development involve large numbers of genes at various developmental stages, and mutations of any of the many regulatory genes could trigger abortion or loss of function of male and/or female organs (Wellmer et al., 2004). In dioecious flowers, male sterile mutation can occur from the early stages of the first appearance of the stamen primordia until the late stages of microspore formation; likewise, female sterility can also occur from early stages of inception of the carpel until the late stage of the formation of microspores (Matsunaga and Kawano, 2001).

Unisexual species have sex determination mechanisms ranging from genetic to environmental, but genetic factors play far greater roles. Genetic mechanisms range from a single locus on an autosome to heteromorphic sex chromosomes containing multiple genes involved in sex determination. In some unisexual species, such as hemp, cucumber, and maize, male or female sterility is influenced by environmental factors such as light intensity, day length, temperature, mineral nutrition, and plant hormones (Frankel and Galun, 1977; Chailakhyan, 1979). Cytoplasmic male sterility caused by mutations on the mitochondrial genome is widely used in production of crops that benefit greatly from heterosis, but only nuclear male sterility can trigger sex chromosome evolution.

Chromosomal sex determination systems may be an evolutionary consequence of natural selection in favor of dioecy. Sex chromosomes do not appear suddenly in animals or plants. Rather, it is the pair of autosomes bearing the sex determination genes that have evolved specialized features, i.e., the degeneration of the Y chromosome, resulting in heteromorphy that became the hallmark of sex chromosomes. Two major sex chromosome systems have evolved. One is the active Y chromosome system, or the XY system (or ZW in a heterogametic female system), in which females have two of the same kind of sex chromosome (XX), while males have two distinct sex chromosomes (XY). The other is the X-to-autosome (A) balance system, in which the ratio of X:A chromosomes determines sex by an X chromosome counting system and the Y chromosome is dispensable.

Among the many dioecious plant species, only a few have evolved sex chromosomes. As in mammals, some dioecious flowering species have an active-Y system of sex determination with heterogametic males (XY) and homogametic females (XX), such as white campion (*Silene latifolia*), papaya (*Carica papaya*), and asparagus (*Asparagus officianalis*). Several dioecious species have the X: A dosage compensation system for sex determination, for example, sorrel (*Rumex*) and hops (*Humulus*). An X-to-autosome ratio of 1.0 or higher results in a female and 0.5 or lower defines a male (Ono, 1935; Westergaard, 1958), which is similar to the situation in Drosophila and Caenorhabditis elegans (reviewed by Mittwoch, 1996). These two sex chromosome systems may reflect two different stages of sex chromosome evolution that will be discussed further in the following sections.

In some species, sex expression is under epigenetic control, mediated by chromatin modifications of the sex determining regions, including DNA methylation and nucleosomal histone acetylation (reviewed by Vyskot, 1999). In white campion, hypomethylation induced sex reversal of genetically male plants (XY) to form a perfect flower (Finnegan et al., 2000). Sex determination is not the simple hierarchical chain reaction once envisioned but is the result of interactions among a complex network of genes as learned from the study of the sex-determining region Y (Sry) gene pathway that controls human sex determination (Koopman, 1999). The epigenetic influence over sex determination might be the consequence of the regulation of the expression of a gene or genes in this network.

SUPPRESSION OF RECOMBINATION AT THE SEX DETERMINATION LOCUS

Suppression of recombination is a pivotal event in sex chromosome evolution. Without suppression of recombination, the male sterile or female sterile mutations could revert to hermaphroditism. The typical features of young sex chromosomes are suppression of recombination at and around the sex determination locus and moderate degeneration of the male-specific region.

Suppression of recombination in specific chromosomal regions is a widespread phenomenon in sexually reproducing organisms. It is often triggered by rearrangement or modification of DNA sequences. Direct evidence was documented for suppression of recombination within two pericentric inversions on human chromosomes 1 and 8 (Jaarola et al., 1998). Inversions were also the cause of suppression of recombination that led to the degeneration of the Y chromosome in human (Lahn and Page, 1999). Although the contents of recombination-suppressed regions evolve rapidly, the regions themselves appear to remain in the same location over long periods (Bowers et al., 2005). Gradual reduction of recombination rates might account in part for reducing recombination of the X and Y chromosomes in Silene (Nicolas et al., 2005). DNA methylation is another mechanism for suppression of recombination as reported in the fungus Ascobolus immersus (Maloisel and Rossignol, 2006).

Suppression of genetic recombination has been documented for primitive and advanced sex chromosomes (Westergaard, 1958; Ohno, 1967; Ma et al., 2004). The papaya sex locus has been genetically mapped to linkage group 1 (Sondur et al., 1996). In a recently constructed high density map, a total of 225 markers, which is 66% of the 342 markers on linkage group 1, co-segregated with the sex locus, indicating severe suppression of recombination at this region (Ma et al., 2004). A similar result has been shown on a linkage map of the medaka fish. Two large clusters of loci consisting of 19 and 13 markers, respectively, flanked a region of 2.6 cM containing the male-determining region (Naruse et al., 2000). In asparagus, the sex locus is located on linkage group 5 where 20 random markers have been mapped to a total of 55.3 cM, averaging one marker in 2.6 cM of this linkage group (Reamon-Büttner and Jung, 2000; Jamsari et al., 2004). However, nearer the sex locus, four markers, excluding the newly integrated sex-linked STS markers, were mapped in a region of 0.5 cM. Recombination suppression at and near the sex locus was further confirmed by physical mapping of the male-specific region in papaya (Liu et al., 2004) and by genetic mapping using male and female meioses in medaka (Kondo et al., 2001).

In the XY system, suppression of recombination and DNA sequence degeneration occur on the Y chromosome, while X chromosomes in females recombine normally. In addition to the large clusters of sex co-segregating markers, suppression of recombination in the male-specific region of the Y chromosome can be evaluated by comparing X and Y linkage maps. In hops (*Humulus lupulus*), the two markers flanking the male sex determination locus were 3.7 cM apart on the Y chromosome linkage map, while the genetic distance between the same two markers was 14.3 cM on the X chromosome linkage map, an almost four-fold reduction of recombination on the Y chromosome (Seefelder et al., 2000). A similar reduction of genetic distance is found in the male meiotic map of the three-spined stickleback, a fish species with an emerging Y chromosome (Peichel et al, 2004). The sex determination gene of sticklebacks is located at the end of linkage group 19. A pair of markers next to the sex determination gene was 6.4 cM apart based on male meiosis (i.e., the primitive Y chromosome) but 25.7 cM apart based on female meiosis (the X chromosome); again recombination on the Y chromosome was reduced approximately four-fold.

Multiple genes in the suppressed male-specific region of the Y chromosome may be specialized in male-specific functions. In white campion, three dispersed male determining loci were located on the Y chromosome, with two on one arm containing genes controlling carpel suppression and early stamen promoting and one on the other arm conferring late anther fertility (Westergaard, 1958; Farbos et al., 1999; Lardon et al., 1999; Lebel-Hardenack et al., 2002). Because these genes are physically located on distant segments of the Y chromosome, they could potentially segregate, resulting in hermaphroditic and sterile individuals unless recombination between them was suppressed.

The Sex of Offspring Is Determined by Particular Chromosomes

In humans and many other animal species, sex is determined by specific chromosomes. How did researchers discover these so-called sex chromosomes? The path from the initial discovery of sex chromosomes in 1891 to an understanding of their true function was paved by the diligent efforts of multiple scientists over the course of many years. As often happens during a lengthy course of discovery, scientists observed and described sex chromosomes long before they knew their function.

An idea inspired by the "X element"

By the 1880s, scientists had established methods for staining chromosomes so that they could be easily visualized using a simple light microscope. With this staining method, scientists were able to observe cell division and to identify the steps that occurred during both mitosis and meiosis.

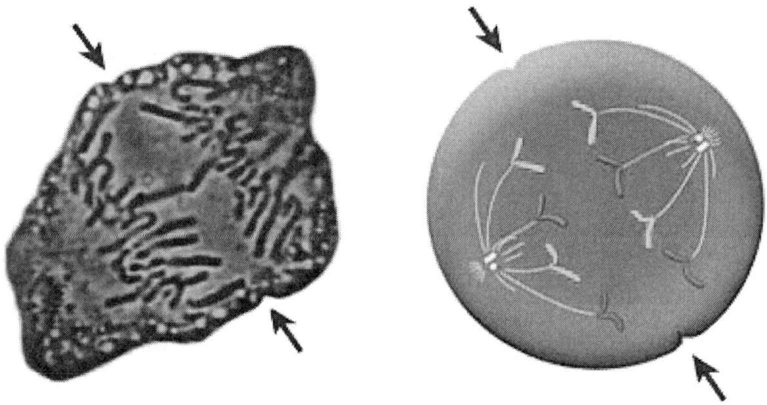

Figure : Cell division observed through the microscope (left) is redrawn to show the action of chromosomes (right). Arrows indicate the axis along which the cell divides.

The first indication that sex chromosomes were distinct from other chromosomes came from experiments conducted by German biologist Hermann Henking in 1891. While using a light microscope to study sperm formation in wasps, Henking noticed that some wasp sperm cells had 12 chromosomes, while others had only 11 chromosomes. Also, during his observation of the stages of meiosis leading up to the formation of these sperm cells, Henking noticed that the mysterious twelfth chromosome looked and behaved differently than the other 11 chromosomes. Accordingly, he named the twelfth chromosome the "X element" to represent its unknown nature. Interestingly, when Henking used a light microscope to study egg formation in female grasshoppers, he was unable to spot the X element.

Based on his observations, Henking hypothesized that this extra chromosome, the X element, must play some role in determining the sex of insects. However, he was unable to gather any direct evidence to support his hypothesis.

Figure : The darkling beetle, *Tenebrio molitor*.

More than a decade after Henking's work, Nettie Stevens surveyed multiple beetle species and examined the inheritance patterns of their chromosomes. In 1905, while studying the gametes of the beetle *Tenebrio molitor*, Stevens noted an unusual-looking pair of chromosomes that separated to form sperm cells in the male beetles. Based on her comparisons of chromosome appearance in cells from male and female beetles, Stevens proposed that these accessory chromosomes were related to the inheritance of sex.

Over time, other scientists studied the appearance of chromosomes in a wide variety of animal species, and it became clear that there was a relationship between the physical appearance and number of chromosomes in gametes and somatic cells from males and females of a given species.

The variety of sex determination systems

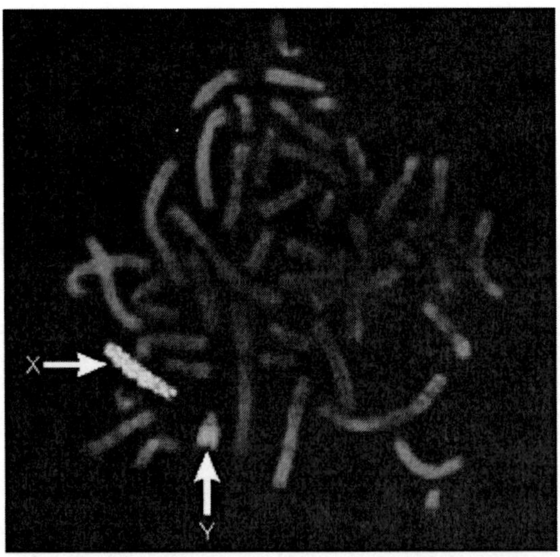

Figure : Example set of male human chromosomes. In the image, the X and Y chromosomes are indicated by arrows.

In humans, females inherit an X chromosome from each parent, whereas males always inherit their X chromosome from their mother and their Y chromosome from their father. Consequently, all of the somatic cells in human females contain two X chromosomes, and all of the somatic cells in human males contain one X and one Y chromosome. The same is true of all other placental mammals — males produce X and Y gametes, and females produce only X gametes. In this system, referred to as the XX-XY system, maleness is determined by sperm cells that carry the Y chromosome.

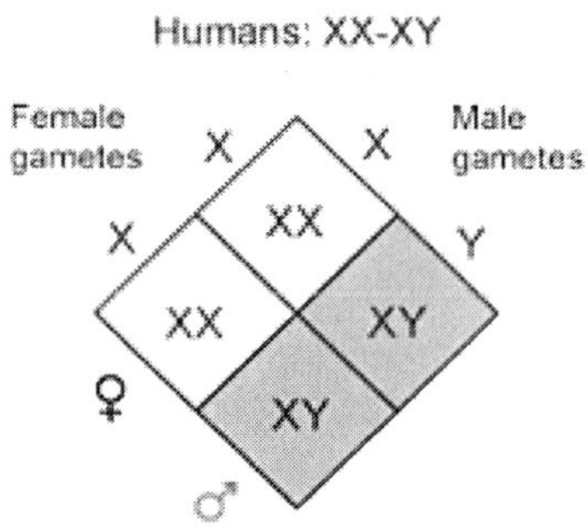

Figure : Sex determination in humans.

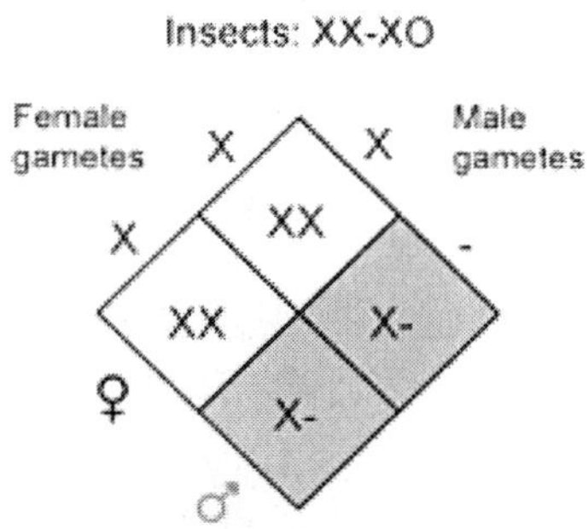

Figure : Sex determination in insects.

Many people do not realize, however, that the XX-XY sex determination system is only one of a variety of such systems within the animal kingdom. In fact, sex determination can be very different between different organisms. For example, in the XX-XO system found in crickets, grasshoppers, and some other insects, sperm cells that lack an X chromosome (referred to as O) determine maleness. Here, females carry two X chromosomes (XX) and only produce gametes with X chromosomes. Males, on the other hand, carry only one X chromosome (XO) and produce some gametes with X chromosomes and some gametes with no sex chromosomes at all.

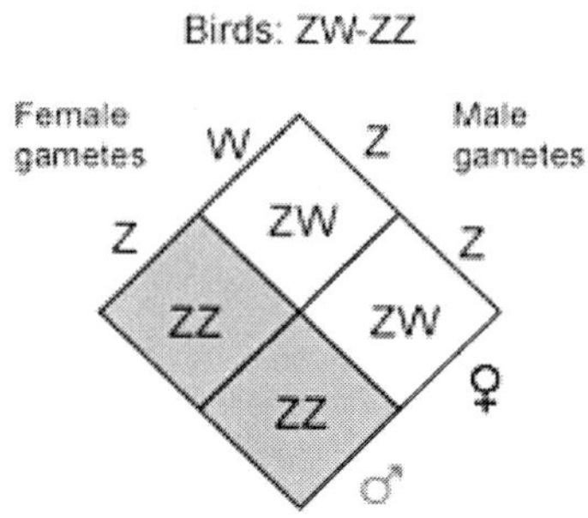

Figure : Sex determination in birds.

Despite the previous examples, males are not always the sex with the mismatched chromosome pair. For example, the ZZ-ZW sex determination system used in birds, snakes, and some insects relies upon females to carry the mismatched chromosome pair (ZW) and males to carry the identical pair (ZZ).

If the three systems discussed above are compared in side-by-side Punnett squares, it is easy to see that sex determination is simply a matter of gamete assortment. Determinations of male and female character arise from a variety of different gamete combination patterns, all of which are the result of gender coding in sexually reproducing organisms.

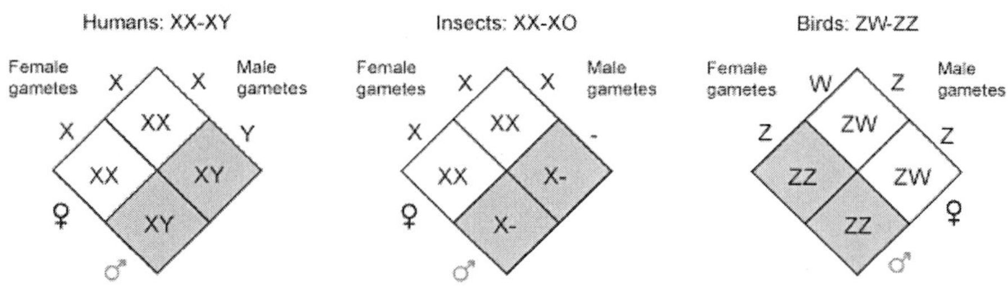

Figure : A side-by-side comparison of sex determination systems in humans, insects, and birds.

The variety of inheritance patterns sex determination is a complex and varied feature among organisms. The XX-XY, XX-XO, and ZZ-ZW systems are only a sample of the wide variety of sex determination systems that scientists have documented in the wide world of living beings, however.

Environmental sex determination – factors temperature control of sex

Environmental sex determination

The sex of the green spoonworm, *Bonellia viridis*, a marine annelid, depends on where the larvae make landfall. (Female shown)

Environmental sex determination is the establishment of sex by a non-genetic cue, such as nutrient availability, experienced within a discrete period after fertilization. Environmental factors which often influence sex determination during development or sexual maturation include light intensity and photoperiod, temperature, nutrient availability, and pheromones emitted by surrounding plants or animals. This is in contrast to genotypic sex determination, which establishes sex at fertilization by genetic factors such as sex chromosomes. Under true environmental sex determination, once sex is determined, it is fixed and cannot be switched again. Environmental sex determination is different from some forms of sequential hermaphroditism in which the sex is determined flexibly after fertilization throughout the organism's life.

Adaptive Significance

Environmental sex determination is similar to certain forms of sexual selection in that there are oftentimes different and opposing selective pressures on males and females because of the costs of reproduction. Sexual selection is common throughout the tree of life (most known in birds); often resulting in sexual dimorphism, or size and appearance differences between sexes in the same species. In environmental sex determination, selective pressures over evolutionary time have selected for flexibility in sex determination to optimize fitness in a heterogenous environment because of the different costs of sex in males and females. Certain environmental conditions differentially affect each sex such that it would be beneficial to become one sex and not the other. This is especially pertinent for sessile organisms that cannot move to a different environment. In plants, for example, female sexual function is often more energetically expensive because once fertilized they must use significant stored energy to produce fruits, seeds, or sporophytes whereas males must only produce sperm (and sperm-containing structure; antheridium in seedless plants, and pollen in seed plants).

Mechanisms

Lacking genetic information coding for separate sexes such as sex chromosomes, individuals that exhibit environmental sex determination contain genetic information coding for both sexes on autosomes. In general, once exposed to certain environmental cues, epigenetic changes cause developing individuals to become either male or female. Environmental cues that often trigger the development of males or females include temperature, nutrient (or food in the case of animals) and water availability, photoperiod, competitive stress, and pheromones from conspecific individuals. Specific mechanisms and cues vary between species.

Taxonomic range

Crustaceans

The amphipod crustacean *Gammarus duebeni* produces males early in thc mating season, and females later, in response to the length of daylight, the photoperiod. Because male fitness improves more than female fitness with increased size, environmental sex determination is adaptive in this system by permitting males to experience a longer growing season than females.

The branchiopod crustacean *Daphnia magna* parthenogenetically produces male progeny in response to a combination of three environmental factors, namely a reduced photoperiod in autumn, shortage of food and raised population density.

Annelids

Bonellia viridis, a marine worm, has location-dependent sex determination; sex depends on where the larvae land.

Vertebrates

The sex of alligators is determined by nest temperature.

The sex of most amniote vertebrates, such as mammals and birds, is determined genetically. However, some reptiles have temperature-dependent sex determination, where sex is permanently determined by thermal conditions experienced during the middle third of embryonic development. The sex of crocodilians and sphenodontians is exclusively determined by temperature. In contrast, squamates (lizards and snakes) and turtles exhibit both genotypic sex determination and temperature-dependent sex determination, although temperature dependence is much more common in turtles than in squamates.

Ferns

A developing fern gametophyte. While in this immature, asexual stage, a fern gametophyte may undergo Environmental Sex Determination based on cues from the surrounding environment.

Most fern species (with a few exceptions, namely the Salvineales) are homosporous and lack sex chromosomes. Lacking genetic information coding for separate sexes, every fern spore has the capacity to become a male, female, or hermaphroditic gametophyte depending on the environment. In many fern species, including *Ceratopteris richardii*, environmental sex determination is linked to breeding systems. Fern gametophytes exhibit a wide variety of breed systems which can be divided into outcrossing and inbreeding. To promote outcrossing, female gametophytes release a chemical pheromone known as Antheridiogen which controls the sex of nearby developing gametophytes. Antheridiogen secreted by females promotes the development of nearby asexual gametophytes into males. This is adaptive because inducing maleness increases the probability of outcrossing as males provide sperm for the females rather than the females becoming hermaphroditic (or bisexual) and self-fertilizing. However, if no fertilization occurs, the female gametophyte can still become hermaphroditic and self-fertilize if the conditions are conducive to growth, ultimately resulting in inbreeding depression.

Additionally, similar to crocodilians, homosporous fern gametophyte sex is determined by the abiotic environment in accordance with the size-advantage model. In stressful environments (crowding or nutrient stress), gametophytes are smaller and develop into males. While in more favorable growing conditions, gametophytes are larger and develop into females.

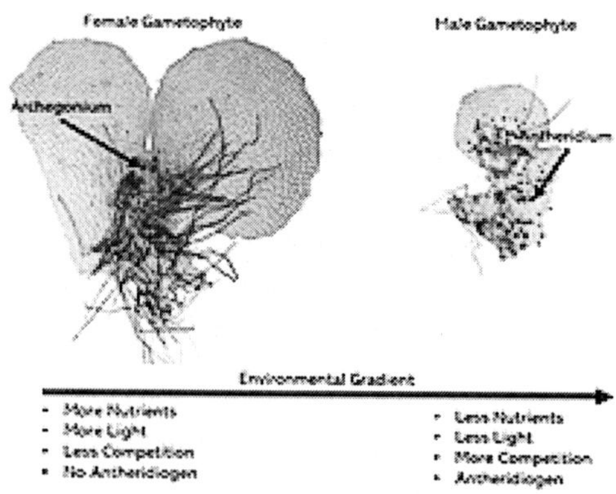

Environmental sex determination in ferns based on nutrient availability, light, competition, and the pheromone Antheridiogen.

Moss

Moss gametophytes (green photosynthetic tissue) and sporophytes (upright, brown structures). The presence of many sporophytes, indicates that the moss is likely either monoicous and self fertilizing or dioicous with fertile males and females. This can be rare as females often outcompete males in mosses.

Moss gametophytes can be either asexual, female, male, or hermaphroditic like ferns. Unlike homosporous ferns, moss gametophytes can be either monoicous or dioicous (similar to monoecious and dioecious in vascular plants), with most studied dioicous species exhibiting genetic sex determination via the UV sex chromosome sex determination system. Some monoicous moss species such as *Splachnum ampullaceum* exhibit environmental sex determination during early development, with low light, low pH, and low nutrient availability all promoting male development. In the presence of auxin, a widespread plant hormone, or gibberellins, compounds similar to Antheridiogen in ferns, both female and male individuals invest more in sexual structures (antheridia and archegonia). Environmental sex determination in moss is fundamentally different from the spatial segregation of sexes, the occurrence of environmentally mediated sex ratios in moss patches, observed in sexually *static* moss species. Spatial segregation of the sexes in mosses is caused by differential *survival* rates between sexes as a result of the competitive advantage of female moss. This leads to female dominated populations maintained by asexual reproduction and minimal sexual reproduction. In contrast, environmental sex determination is the *dynamic development* of females or males in different environmental conditions.

Angiosperms

Flowering catasetum viridiflavum.

Many angiosperms exhibit sequential hermaphroditism, meaning that they can switch sexes continually throughout their life based on the current conditions and resource availability to optimize fitness each flowering season. But sequential hermaphroditism and environmental sex determination are not mutually exclusive. For example, Catasetum viridiflavum, an epiphyte (plant that grows on another plant) in the Orchidaceae family exhibits sequential hermaphroditism where the younger, smaller individuals have male inflorescences and the older, larger individuals have female inflorescences, but sex expression is also strongly influenced by light intensity. Individuals in high light are more often female and individuals in the low light are more often male, regardless of size. In higher light, individuals produce more ethylene, a common plant hormone, which promotes the formation of female flowers.

Hormonal control: dosage compensation in organisms with inheritance genetic Males heterogametic females

The complex interplay between the glands, hormones and other target organs is referred to as the endocrine system. The endocrine system is one of two main communication systems in the body together with the nervous system.

Endocrine glands are small organs that make and release hormones directly into the bloodstream.

At its simplest a hormone is a chemical messenger from one cell or group of cells to another. Hormones are released (secreted) into the bloodstream and have an effect on other parts of the body.

Hormones are found in all organisms with more than one cell, and so they are found in plants and animals. They influence or control a wide range of physiological activities such as growth, development, puberty, regulating sugar levels, bone growth and appetite.

Sex-chromosome dosage compensation

Dosage compensation is the process by which organisms equalize the expression of genes between members of different biological sexes. Across species, different sexes are often characterized by different types and numbers of sex chromosomes. In order to neutralize the large difference in gene dosage produced by differing numbers of sex chromosomes among the sexes, various evolutionary branches have acquired various methods to equalize gene expression among the sexes. Because sex chromosomes contain different numbers of genes, different species of organisms have developed different mechanisms to cope with this inequality. Replicating the actual *gene* is impossible; thus organisms instead equalize the *expression* from each gene. For example, in humans, female (XX) cells randomly silence the transcription of one X chromosome, and transcribe all information from the other, expressed X chromosome. Thus, human females have the same number of expressed X-linked genes per cell as do human males (XY), both sexes having essentially one X chromosome per cell, from which to transcribe and express genes.

Different lineages have evolved different mechanisms to cope with the differences in gene copy numbers between the sexes that are observed on sex chromosomes. Some lineages have evolved dosage compensation, an epigenetic mechanism which restores expression of X or Z specific genes in the heterogametic sex to the same levels observed in the ancestor prior to the evolution of the sex chromosome. Other lineages equalize the expression of the X- or Z- specific genes between the sexes, but not to the ancestral levels, i.e. they possess incomplete compensation with "dosage balance". One example of this is X-inactivation which occurs in humans. The third documented type of gene dose regulatory mechanism is incomplete compensation without balance (sometimes referred to as incomplete or partial dosage compensation). In this system gene expression of sex-specific loci is reduced in the heterogametic sex i.e. the females in ZZ/ZW systems and males in XX/XY systems.

There are three main mechanisms of achieving dosage compensation which are widely documented in the literature and which are common to most species. These include random inactivation of one female X chromosome (as observed in humans and *Mus musculus*; this is called X-inactivation), a two-fold increase in the transcription of a single male X chromosome (as observed in *Drosophila melanogaster*), and decreased transcription by half in both of the X chromosomes of a hermaphroditic organism (as observed in *Caenorhabditis elegans*). These mechanisms have been widely studied and manipulated in model organisms commonly used in the laboratory research setting. A summary of these forms of dosage compensation is illustrated below. However, there are also other less common forms of dosage compensation, which are not as widely researched and are sometimes specific to only one species (as observed in certain bird and monotreme species).

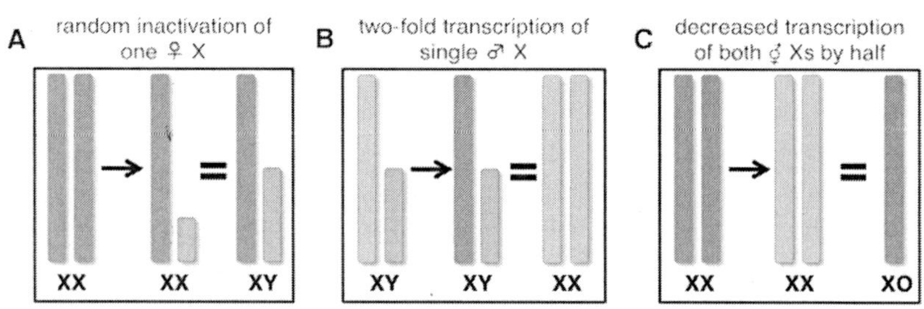

Figure 1. Three main mechanisms of dosage compensation observed in common model eukaryotic organisms.
(A) Mus musculus (as well as humans and many mammalian organisms) follow a pattern of dosage compensation in which females randomly inactive one X chromosome in every cell in their body to match male expression of a single X chromosome.
(B) D. melanogaster follow a mechanism of dosage compensation in which males increase the transcription of their single X chromosome by two-fold in order to match female expression of two X chromosomes.
(C) C. elegans hermaphroditic worms express both X chromosomes by half to match male worms with only one X chromosome.

Three main mechanisms of dosage compensation observed in common model eukaryotic organisms.

Random inactivation of one @& X

One logical way to equalize gene expression amongst males and females that follow a XX/XY sex differentiation scheme would be to decrease or altogether eliminate the expression of one of the X chromosomes in an XX, or female, homogametic individual, such that both males and females then express only one X chromosome. This is the case in many mammalian organisms, including humans and mice.

The evidence for this mechanism of dosage compensation was discovered prior to scientists' understanding of what its implications were. In 1949, Murray Barr and Ewert Bertram published data describing the presence of "nucleolar satellites, which they observed were present in the mature somatic tissue of different female species. Further characterization of these satellites revealed that they were actually packages of condensed heterochromatin, but a decade would pass before scientists grasped the significance of this specialized DNA.

Then, in 1959 Susumu Ohno proved that these satellite-like structures found exclusively in female cells were actually derived from female X chromosomes. He called these structures Barr bodies after one of the investigators who originally documented their existence. Ohno's studies of Barr bodies in female mammals with multiple X chromosomes revealed that such females used Barr bodies to inactivate all but one of their X chromosomes. Thus, Ohno described the "n-1" rule to predict the number of Barr bodies in a female with n number of X chromosomes in her karyotype.

Simultaneously, Mary F. Lyon began investigating manipulations of X-linked traits that had phenotypically visible consequences, particularly in mice, whose fur color is a trait intimately linked to the X chromosome. Building on work done by Ohno and his colleagues, Lyon eventually proved that either the maternal or paternal X chromosome is randomly inactivated in every cell of the female body in the species she was studying, which explained the heterogeneous fur patterns she observed in her mosaic mice. This process is known as X-inactivation, and is sometimes referred to as "lyonization". This discovery can be easily extrapolated to explain the mixed color patterns observed in the coats of tortoiseshell cats. The fur patterns characteristic of tortoiseshell cats are found almost exclusively in females, because only they randomly inactivate one X chromosome in every somatic hair cell. Thus, presuming that hair color determining genes are X-linked, it makes sense that whether the maternal or paternal X chromosome is inactivated in a particular hair cell can result in differential fur color expression.

Compounding on Lyon's discoveries, in 1962 Ernest Beutler used female fibroblast cell lineages grown in culture to demonstrate the heritability of lyonization or random X-inactivation. By analyzing the differential expression of two existing, viable alleles for the X-linked enzyme glucose-6-phosphate dehydrogenase (G6PD) gene, Beutler observed that the inactivation of the gene was heritable across passaged generations of the cells.

This pattern of dosage compensation, caused by random X-inactivation, is regulated across development in female mammals, following concerted patterns throughout development; for example, at the beginning of most female mammal development, both X chromosomes are initially expressed, but gradually undergo epigenetic processes to eventually achieve random inactivation of one X. In germ cells, inactivated X chromosomes are then once again activated to ensure their expression in gametes produced by female mammals.

Thus, dosage compensation in mammals is largely achieved through the silencing of one of two female X chromosomes via X-inactivation. This process involves histone tail modifications, DNA methylation patterns, and reorganization of large-scale chromatin structure encoded by the X-ist gene. In spite of these extensive modifications, not all genes along the X chromosome are subject to X-inactivation; active expression at some loci is required for homologous recombination with the pseudo-autosomal region (PAR) of the Y chromosome during meiosis. Additionally, 10-25% of human X chromosome genes, and 3-7% of mouse X chromosome genes outside of the PARs show weak expression from the inactive X chromosome.

Random X-inactivation demands that the cell can determine if it contains more than one active X-chromosome before acting to silence any extraneous X-chromosome(s). This process is known as "counting". The exact molecular mechanism of counting is still unknown, but a popular model posits that autosomes produce factors that repress X-inactivation, while X-chromosome products that promote X-inactivation. These two conflicting forces are balanced such that if there is more than one X-chromosome X-inactivation will occur, but if there is only one, the autosomal products will successfully prevent the process.

Not all random X-inactivation is entirely random. Some alleles, generally mutations in the X-inactivation center on the X-chromosome have been demonstrated to confer a bias towards inactivation for the chromosome on which they sit. Truly random X-inactivation may also appear to be non-random if one X-chromosome carries a deleterious mutation. This can result in fewer cells which express the lower-fitness X-chromosome to be present in the body as these cells are selected against.

Two-fold increased transcription of a single B& X

Another mechanism common for achieving equal X-related genetic expression between males and females involves two-fold increased transcription of a single male X chromosome. Thus, heterogametic male organisms with one X chromosome may match the level of expression achieved in homogametic females with two active X chromosomes. This mechanism is observed in *Drosophila.*

The concept of dosage compensation actually originated from an understanding of organisms in which males upregulated X-linked genes two-fold, and was much later extended to account for the observation of the once mysterious Barr bodies. As early as 1932, H.J. Muller carried out a set of experiments which allowed him to track the expression of eye color in flies, which is an X-linked gene. Muller introduced a mutant gene that caused loss of pigmentation in fly eyes, and subsequently noted that males with only one copy of the mutant gene had similar pigmentation to females with two copies of the mutant gene. This led Muller to coin the phrase “dosage compensation” to describe the observed phenomenon of gene expression equalization.

Despite these advances, it was not until Ardhendu Mukherjee and W. Beermann performed more advanced autoradiography experiments in 1965 that scientists could confirm that transcription of genes in the single male X chromosome was double that observed in the two female X chromosomes. Mukherjee and Beermann confirmed this by designing a cellular autoradiography experiment that allowed them to visualize incorporation of [3H]uridine into ribonucleic acid of the X chromosomes. Their studies showed equal levels of [3H]uridine incorporation in the single male X chromosome and the two female X chromosomes. Thus, the investigators concluded that the two-fold increase in the rate of RNA synthesis in the X chromosome of the male relative to those of the female could account for Muller’s hypothesized dosage compensation.

In the case of two-fold increased transcription of a single male X chromosome, there is no use for a Barr body, and the male organism must use different genetic machinery to increase the transcriptional output of their single X chromosome. It is common in such organisms for the Y chromosome to be necessary for male fertility, but not for it to play an explicit role in sex determination. In *Drosophila*, for example, the sex lethal (SXL) gene acts as a key regulator of sexual differentiation and maturation in somatic tissue; in XX animals, SXL is activated to repress increased transcription, while in XY animals SXL is inactive and allows male development to proceed via increased transcription of the single X. Several binding sites exist on the *Drosophila* X chromosome for the dosage compensation complex (DCC), a ribonucleoprotein complex; these binding sites have varying levels of affinity, presumably for varying expression of specific genes. The Male Specific Lethal complex, composed of protein and RNA binds and selectively modifies hundreds of X-linked genes, increasing their transcription to levels comparable to female *D. melanogaster*.

In organisms that use this method of dosage compensation, the presence of one or more X chromosomes must be detected early on in development, as failure to initiate the appropriate dosage compensation mechanisms is lethal. Male specific lethal proteins (MSLs) are a family of four proteins that bind to the X chromosome exclusively in males. The name "MSL" is used because mutations in these genes cause inability to effectively upregulate X-linked genes appropriately, and are thus lethal to males only and not their female counterparts. SXL regulates pre-messenger RNA in males to differentially splice MSLs and result in the appropriate increase in X chromosome transcription observed in male *Drosophila*. The immediate target of SXL is male specific lethal-2 (MSL-2). Current dogma suggests that the binding of MSL-2 at multiple sites along the SXL gene in females prevents proper MSL-2 translation, and thus, as previously stated, represses the possibility for X-linked genetic upregulation in females. However, all other transcription factors in the MSL family—maleless, MSL-1, and MSL-3—are able to act when SXL is not expressed, as in the case in males. These factors act to increase male X chromosome transcriptional activity. Histone acetylation and the consequent upregulation of X-linked genes in males is dictated by the MSL complex. Specifically, special roX non-coding RNAs on the MSL complexes facilitate binding to the single male X chromosome, and dictate acetylation of specific loci along the X chromosome as well as the formation of euchromatin. Though these RNAs bind at specific sites along the male X chromosome, their effects spread along the length of the chromosome and have the ability to influence large-scale chromatin modifications. The implications of this spreading epigenetic regulation along the male X chromosome is thought to have implications for understanding the transfer of epigenetic activity along long genomic stretches.

Decreased transcription of both hermaphroditic Xs by half

Other species that do not follow the previously discussed conventions of XX females and XY males must find alternative ways to equalize X-linked gene expression among differing sexes. For example, in *Caenorhabditis elegans* (or *C. elegans*), sex is determined by the ratio of X chromosomes relative to autosomes; worms with two X chromosomes (XX worms) develop as hermaphrodites, whereas those with only one X chromosome (XO worms) develop as males. This system of sex determination is unique, because there is no male specific chromosome, as is the case in XX/XY sex determination systems. However, as is the case with the previously discussed mechanisms of dosage compensation, failure to express X-linked genes appropriately can still be lethal.

In this XX/XO sex determination system, gene expression on the X chromosome is equalized by downregulating expression of genes on both X chromosomes of hermaphroditic XX organisms by half. In these XX organisms, the dosage compensation complex (DCC) is assembled on both X chromosomes to allow for this tightly regulated change in transcription levels. The DCC is often compared to the condensin complex, which is conserved across the mitotic and meiotic processes of many species. This complex is crucial to the condensation and segregation of chromosomes during both meiosis and mitosis. Because data substantiates the theory that dosage compensation in other species is caused by chromatin-wide modifications, many theorize that the DCC in particular functions similar to the condensin complex in its ability to condense or remodel the chromatin of the X chromosome.

The role of the DCC in this form of dosage compensation was postulated by Barbara J. Meyer in the 1980s, and its individual components and their cooperative function were later parsed out by her lab. Notably, in 1999, data from Meyer's lab showed that SDC-2 is a particularly important transcriptional factor for targeting the DCC to the X chromosome and for assembling DCC components onto the X chromosomes in XX embryos. More recently, Meyer's lab has shown that proteins known as X-linked signal elements (XSEs) operate in concert with SDC-2 to differentially repress and activate other genes in the dosage compensation pathway. By selectively mutating a panel of genes hypothesized to contribute to dosage compensation in worms, Meyer's group demonstrated which XSEs specifically play a role in determining normal dosage compensation. They found that during embryonic development, several X-linked genes—including sex-1, sex-2, fox-1, and ceh-39—act in a combinatorial fashion to selectively repress transcriptional activity of the xol-1 gene in hermaphrodites. Xol-1 expression is tightly regulated during early development, and is considered the most upstream gene in sex determination of C. elegans. In fact, xol-1 is often referred to in the literature as the master sex regulatory gene of C. elegans. XX C. elegans embryos have much lower xol-1 expression than their XO counterparts, resulting from overall increases in the amount of SEX-1, SEX-2, CEH-39, and FOX-1 transcription produced in the female embryos. This consequent decrease in xol-1 expression then allows higher SDC-2 expression levels, which aids in the formation and function of the DCC complex in the XX hermaphroditic worms, and in turn results in equalized expression of X-linked genes in the hermaphrodite.

Though all of the above-mentioned XSEs act to reduce xol-1 expression, experimentally reducing expression levels of these individual XSEs has been shown to have a minimal effect on sex determination and successful dosage compensation. This could be in part because these genes encode different proteins that act cooperatively rather than in an isolated fashion; for example, SEX-1 is a nuclear hormone receptor, while FOX-1 is an RNA-binding protein with properties capable of inducing post-transcriptional modifications in the xol-1 target. However, reducing the level of more than one XSE in different combinational permutations seems to have an additive effect on ensuring proper sex determination and resultant dosage compensation mechanics. This supports the hypothesis that these XSEs act together to achieve the desired sex determination and dosage compensation fate. Thus, in this model organism, the achieved level of X-chromosome expression is directly correlated to the activation of multiple XSEs that ultimately function to repress xol-1 expression in a developing worm embryo. A summary of this *C. elegans* mechanism of dosage compensation is illustrated below.

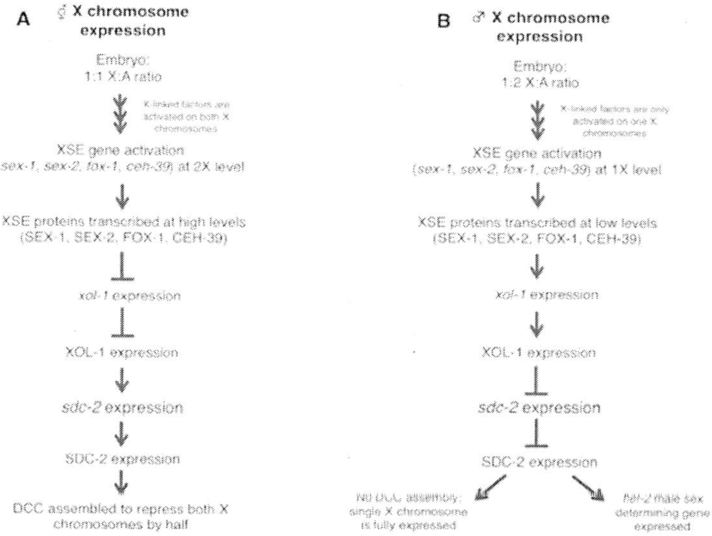

Figure 2. Mechanisms for differential hermaphrodite and male X chromosome expression in *C. elegans*.

(A) Hermaphroditic worms have two X chromosomes and thus initiate the dosage compensation signaling cascade by expressing twice the level of XSEs that male worms are able to. This initiates a pathway that results in assembly of the DCC, providing the machinery for decreased X chromosome expression (dosage compensation)..

(B) Male embryonic worms have only one X chromosome, and can thus only express half the XSEs expressed by female worms. They ultimately do not activate genes that form the DCC, and instead express male sex-determining genes to allow full expression of all genes on their single X chromosome.

Other species-specific methods

The **ZZ/ZW sex system** is used by most birds, as well as some reptiles and insects. In this system the Z is the larger chromosome so the males (ZZ) must silence some genetic material to compensate for the female's (ZW) smaller W chromosome. Instead of silencing the entire chromosome as humans do, male chickens (the model ZZ organism) seem to engage in selective Z silencing, in which they silence only certain genes on the extra Z chromosome. Thus, male chickens express an average of 1.4-1.6 of the Z chromosome DNA expressed by female chickens. The Z chromosome expression of male zebra finches and chickens is higher than the autosomal expression rates, whereas X chromosome expression in female humans is equal to autosomal expression rates, illustrating clearly that both male chickens and male zebra finches practice incomplete silencing. Few other ZZ/ZW Systems have been analyzed as thoroughly as the chicken; however a recent study on silkworms revealed similar levels of unequal compensation across male Z chromosomes. Z-specific genes were over-expressed in males when compared to females, and a few genes had equal expression in both male and female Z chromosomes.

In chickens, most of the dosage compensated genes exist on the Zp, or short, arm of the chromosome while the non-compensated genes are on the Zq, or long, arm of the chromosome. The compensated (silenced) genes on Zp resemble a region on the primitive platypus sex chromosome, suggesting an ancestor to the XX/XY system.

Birds

The sex chromosomes of birds evolved separately from those of mammals and share very little sequence homology with the XY chromosomes. As such, scientists refer to bird sex chromosomes as a ZW sex-determining system, with males possessing two Z chromosomes, and females possessing one Z chromosome and one W. Thus, dosage compensation in birds could be hypothesized to follow a pattern similar to the random X-inactivation observed in most mammals. Alternatively, birds might show decreased transcription of the two Z chromosomes present in the male heterogametic sex, similar to the system observed in the two hermaphrodite X chromosomes of *C. elegans*. However, bird mechanisms of dosage compensation differ significantly from these precedents. Instead, male birds appear to selectively silence only a few genes along one of their Z chromosomes, rather than randomly silencing one entire Z chromosome. This type of selective silencing has led some people to label birds as "less effective" at dosage compensation than mammals. However, more recent studies have shown that those genes on the Z chromosome which are not inactivated in birds may play an important role in recruiting dosage compensation machinery to the Z chromosome in ZZ organisms. In particular, one of these genes, ScII has been demonstrated to be an ortholog of xol-1, the master sex regulator gene in C. elegans. Thus, the function of the selective silencing may be to spare dosage compensation of genes crucial for sex determination of homologous pairing.

While the epigenetic mechanisms behind dosage compensation in birds are poorly understood, especially in comparison to the well-studied mechanisms of dosage compensation in humans and *Drosophila*, several recent studies have revealed promising sequences. One example is MHM (male hypermethylated) RNA, an Xist-like long noncoding RNA that is expressed only in female chickens (ZW). It is associated with female-specific hyper-acetylation of lysine 16 on histone 4 near the MHM locus on the Z chromosome. This MHM locus is heavily studied as a site of dosage compensation because male Z chromosomes are hypermethylated and thus underexpress genes in this area in comparison to female Z chromosomes which are hyperacetylated and overexpress these genes. There has been debate on whether the MHM locus constitutes dosage compensation, however, since scientists claim that even if the MHM locus has been found to have significantly greater expression in females than in males, it could not even be considered to be a dosage compensation mechanism since it does not balance gene dose between the Z chromosome and autosomes in the heterogametic sex.

Similar to mammals, chickens seem to use CpG islands (segments of Cytosine-phosphate-Guanine that are more readily methylated and silenced than other DNA segments) to regulate gene expression. One study found that CpG islands were found primarily in compensated areas of the Z chromosome—areas that are differentially expressed in male and female chickens. Thus it is likely that these CpG islands are locations where genes on the male Z chromosome are methylated and silenced, but which stay functional on the female Z chromosome.

Monotremes

Platypus, a type of monotreme

Monotremes are a class of basal mammals that also lay eggs. They are an order of mammals that includes platypuses and four species of echidna, all of which are egg-laying mammals. While monotremes use an XX/XY system, unlike other mammals, monotremes have more than two sex chromosomes. The male short-beaked echidna, for example, has nine sex chromosomes—5 Xs and 4 Ys, and the male platypus has 5 Xs and 5 Ys.

Platypuses are a monotreme species whose mechanism of sex determination has been extensively studied. There is some contention in academia about the evolutionary origin and the proper taxonomy of platypuses. A recent study revealed that four platypus X chromosomes, as well as a Y chromosome, are homologous to some regions on the avian Z chromosome. Specifically, platypus X1 shares homology with the chicken Z chromosome, and both share homology with the human chromosome 9. This homology is important when considering the mechanism of dosage compensation in monotremes. In 50% of female platypus cells, only one of the alleles on these X chromosomes is expressed while in the remaining 50% multiple alleles are expressed. This, combined with the portions that are homologous to chicken Z and human 9 chromosomes imply that this level of incomplete silencing may be the ancestral form of dosage compensation.

Regardless of their ambiguous evolutionary history, platypuses have been empirically determined to follow an XY sex-determination system, with females possessing five pairs of X chromosomes as the homogametic sex, and males possessing five X and five Y chromosomes as the heterogametic sex. Because the entire genome of the platypus has yet to be completely sequenced (including one of the X chromosomes), there is still continued investigation as to the definitive mechanism of dosage compensation that Platypuses follow. Research from the laboratory of Jennifer Graves used qPCR and SNP analysis of BACs containing various genes from X chromosomes in order to find whether multiple alleles for particular X-linked genes were being expressed at once, or were otherwise being dosage compensated. Her group found that in female platypuses, some X-linked genes only expressed an allele from one X chromosomes, while other genes expressed multiple alleles. This appears to be a system similar to the selective silencing method of dosage compensation observed in birds. However, about half of all X-linked genes also seemed to stochastically express only one active copy of said gene, alluding to the system of random X-inactivation observed in humans. These findings suggest that platypuses may employ a hybrid form of dosage compensation that combines feature from mammals as well as birds. Understanding the evolution of such a system may have implications for solidifying the true ancestral lineage of monotremes.

Plants

In addition to humans and flies, some plants also make use of the XX/XY dosage compensation systems. *Silene latifolia* plants are also either male (XY) or female (XX), with the Y chromosome being smaller, with fewer genes expressed, than the X chromosome. Two separate studies have shown male *S. latifolia* expression of X-linked genes to be about 70% of the expression in females. If the *S. latifolia* did not practice dosage compensation, the expected level of X-linked gene expression in males would be 50% that of females, thus the plant practices some degree of dosage compensation but, because male expression is not 100% that of females, it has been suggested that *S. latiforia* and its dosage compensation system is still evolving. Additionally, in plant species that lack dimorphic sex chromosomes, dosage compensation can occur when aberrant meiotic events or mutations result in either aneuploidy or polyploidy. Genes on the affected chromosome may be upregulated or down-regulated to compensate for the change in the normal number of chromosomes present.

Reptiles

Research into dosage compensation has been carried out in six species of toxicoferan reptiles and in one species of softshell turtle. Two species of caenophidian snake (one which belongs to the family Viperidae and the other to the family Colubridae) have been investigated and both of these exhibit female heterogametic sex determination systems (ZZ\ZW) and have incomplete compensation without balance. The Komodo dragon exhibits incomplete compensation without dosage balance in their independently evolved ZZ/ZW system. In the XX/XY system of *Basiliscus vittatus* and multiple neo-sex chromosomes with male heterogamety in the pygopodid gecko *Lialis burtonis* incomplete compensation without dosage balance were also seen. The Green anole (*Anolis carolinensis*; Dactyloidea), has XX/XY sex determination and unlike the other squamates studied to date has complete dosage compensation with dosage balance. In the Florida softshell turtle (*Apalone ferox)* with ZZ/ZW sex chromosomes, the lack of dosage balance in the expression of Z-linked genes was also found.

X chromosome inactivation and embryonic stem cells

XCI is initiated very early during female embryonic development or upon differentiation of female embryonic stem (ES) cells and results in inactivation of one X chromosome in every female somatic cell. This process is initiated very early during development, around the two- to eight-cell stage and is maintained in the developing extra-embryonic tissues of the embryo, including the fetal placenta. Xist RNA induces heterochromatinization of the X chromosome by attracting chromatin modifiers, involved in gene silencing. Xist RNA is tightly associated with the Xi and it is required for X Chromosome Inactivation to occur in cis. Knockout studies in female ES cells and mice have shown that X chromosomes bearing a deletion of the Xist gene are unable to inactivate the mutated X. Most of the human female ES cell lines display an inactivated X chromosome already in the undifferentiated state characterized by XIST expression, XIST coating and accumulated markers of heterochromatin on the Xi.

It is widely thought that human embryos do not employ XCI prior to implantation. Female embryos have an accumulation of Xist RNA on one of the two X chromosomes, beginning around the 8-cell stage. Xist RNA accumulates at the morula and blastocyst stages and is shown to be associated with transcriptional silencing of the Xist-coated chromosomal region, therefore indicating dosage compensation has occurred. Recently, however, it has become increasingly apparent that XCI of the paternal X chromosome is already present from the 4-cell stage onward in all cells of preimplantation mouse embryos, not the 8-cell stages.

Xist, Xite, and Tsix and their roles in X-inactivation

Xite and Xist, are both long non-coding RNAs that regulate and facilitate the process of X-inactivation and are important in the silencing of genes within the X chromosome that is being inactivated. These work in combination with Tsix, which is non-coding RNA that is an antisense which downregulates the effects of Xist on the X chromosome in which it is expressed on the maternal X chromosome upon the start regulation of X-inactivation. These three RNAs regulate the X-X pair in a *cis*orientation in order to be able to have both chromosomes available for inhibitory actions. Tsix and Xite have basic lncRNAs functions in addition to X-inactivation and regulate the X-X pair in the *trans*orientation. This ensures exclusive silencing for both X chromosomes. Xite and Tsix are both essential within the orientational directional processes in *cis* and *trans* as it is seen that without Tsix and Xite in trans it perturbs pairing and counting of genes.

Once the Xist is turned off and no longer regulates the process, the Tsix will slowly decrease in expression as well until both RNAs are no longer being changed by Xic. Xite is the locus that harbors intergenic transcription start sites from hypersensitive sites of allelic crossovers/ differences. When X-inactivation begins, the transcription of Xite increases and signals for the downregulation of Tsix in *cis*orientation, which is on the silent X chromosome, all while promoting the Tsix persistence on the active X chromosome. Xite also has major roles to play in the asymmetry of Tsix expression and generates X chromosome inequality through moving and helping orient the chromosomes to be acted upon by the correct subsequent lncRNA, either Tsix or Xist.

Neo-sex chromosomes and Dosage compensation

The monarch butterfly *Danaus plexippus* belongs to the order Lepidoptera and has 30 chromosomes one of which is a neo-sex chromosome which is the result of a fusion between one of the sex chromosomes and an autosome. A study using a combination of methods (Hi-C assembly, coverage analysis and ChIp-seq) found that the neo-Z segment exhibits complete dosage compensation which is achieved by increased transcription in ZW females. Interestingly, the ancestral Z segment exhibits dosage balance with transcription levels being equal between both genders but less than the expected ancestral level, and this is achieved by decreased transcription in ZZ males.

Lack of Dosage Compensation in Organisms with Heterogametic Females

In the above discussion, we noticed that when male sex is heterogametic (XX@&, XYB& or XX@&, XOB&), X-linked genes (more than one hundred genes) are subject to dosage compensation. In contrast to this, when the female sex is heterogametic (ZZ B&, ZW @&), as in birds, butterflies and moths, Z-linked genes are apparently not dosage compensated. Similar situation exists in some reptiles and. amphibians, where female heterogamety is predominant. A study of the absence of dosage compensation in these heterogametic females led to the following conclusions : (i) Genes which, require dosage compensation are primarily those that control morphogenesis and the prospective body plan, (ii) The products of these genes are ' required in disomic doses especially during oogenesis and early embryonic development, (iii) During oogenesis itself, heterogametic females synthesize and store morphogenetically essential gene products, including those encoded by Z-linked genes, in large quantities, (iv) Abundance of these gene products in the egg and their persistence relatively late in embryogenesis enables heterogametic females to overcome the monosomic state of the Z-chromosome in ZW embryos. Whenever the females are heterogametic, they have *megalecithal* eggs containing several thousand times more maternal RNA and other maternal messages than eggs of organisms, where male is the heterogametic sex.

Among amphibians, one example of female heterogametic sex (ZW) is *Xenopus laevis,* an organism extensively utilized for genetic studies. Its egg is 4000 times the size of egg in mouse or human, and it contains 10,000 times as much stored mRNA, proteins, lipids and other constituents. In XX females (e.g., mice, humans, etc.), this need for higher activity of X-linked genes is satisfied by the ' XX chromosome constitution of the zygote. In ZW females, the inadequate gene dosage of Z-linked genes is overcome in two ways : (i) increased synthesis and storage of the products of maternally acting genes; (ii) a shift in the timing of action of zygotic genes, such that they become effectively 'maternal genes' in early stages. These differences between organisms having female heterogamety and those having male heterogamety allow survival of heterogametic females without dosage compensation.

Genetic Imprinting

Genetic imprinting is a phenomenon, which involves differences in expression of genes inherited from mother and father. In other words, the chromosomes in the sperm and egg are differentially imprinted or marked as having come from the father or the mother, so that some maternally inherited genes and other paternally inherited genes are differentially expressed in the progeny.

Any epigenetic changes in the germ cells, such as the inactivation of an X-chromosome, are reversed or altered during or just before meiosis, which leads to the production of sperms and eggs. After this reversal of epigenetic changes, the eggs and sperms are differentially imprinted. As a consequence of this, even if a chromosome in a sperm happens to have been inherited from the fathers' mother, it would still bear the male imprint. Such imprinting has been found to be crucial to normal development in mice and other mammals, because it silences and activates different sets of genes in the maternal and paternal chromosomes. These differences in paternal and maternal chromosomes complement each other in directing normal development. This differentiation or reprogramming of the genome is believed to be achieved through changes in methylation patterns. In mice, differential methylation of paternal and maternal chromosomes has actually been demonstrated.

A case of genetic imprinting has also been shown in human patients suffering with a disease known as Beckwith Wiedermann Syndrome (BWS), where both copies of a region on the short arm of chromosome 11 (*Up 15.5*)were inherited from the father. This feature is described as paternal disomy. A similar situation has been shown involving paternal disomy for mouse chromosome 7 (homologous to human chromosome 11), which led to production of abnormally large embryos, a condition analogous to BWS in humans. This is attributed to differential paternal expression (imprinting) of a gene *Igf*2(insulin like growth factor-2) in mouse, which is believed to be homologous to BWS-gene in humans.

Dosage compassion in plants - problems

We take plants entirely for granted, farming them for food and timber, cultivating selected garden plants for pleasure and chemically killing others ("weeds") that don't please our eye, and so on. We have no concept of plant welfare in the way we have of animal welfare – no concern for the comfort and state of mind of the individual plant.

"That's simple, genius," you may retort, "plants don't have feelings or minds." But, things are not that straightforward. Plants are incredibly successful and sophisticated life forms and may have inner lives. Michael Pollan wrote a fascinating review titled The Intelligent Plant in The New Yorker (December 23rd, 2013).

Plants are easily the dominant life form on Earth, constituting 80 per cent of total biomass – bacteria constitute 13 per cent and fungi 2 per cent. Humanity is only a trace life form on Earth, constituting about one-10,000th of the total biomass.

Biological life is broadly divided into autotrophic and heterotrophic organisms. Plants are autotrophs and animals are heterotrophs. Plants are much more biochemically versatile than animals. Autotrophs can synthesise their own food from simple chemicals available in the environment (CO2, N2, H2O). Heterotrophs cannot do this and must eat other organisms, both plants and animals, for nutrition.

Chapter 9

Maternal Effects and Cytoplasmic Inheritances

Introduction to Cytoplasmic inheritance

Cytoplasmic inheritance is also called as the extranuclear inheritance, maternal inheritance, non-Mendelian inheritance, non-chromosomal inheritance or extra chromosomal inheritance.

Chromosomal DNA is the main storehouse of genetic information in a cell. It is instrumental in deciding the phenotype of an offspring. However, there are instances where the phenotype of the offspring is similar to the maternal phenotype regardless of environmental effects or the genotype it bears. This indicates that there is DNA outside the nucleus which contributes to deciding the phenotype of the offspring. Scientists have discovered that it is mainly due to two phenomena named cytoplasmic inheritance and genetic maternal effect. Though chromosomes divide precisely into gametes during meiosis, the cytoplasm of gametes does not collect precisely into the zygote. Cytoplasmic inheritance and genetic maternal effects arise due to the contribution of more cytoplasm by female gamete into the resultant zygote during the syngamy. However, the cytoplasmic inheritance and genetic maternal effect are different from each other. The key difference between cytoplasmic inheritance and genetic maternal effect is that cytoplasmic inheritance occurs due the genetic information stored in genes of some organelles such as mitochondria and chloroplasts present in the cytoplasm while genetic maternal effect occurs due to the mRNA and proteins received from the female gamete.

Cytoplasmic inheritance - involving hereditary units nuclear genes

Mitochondria and chloroplast are two organelles present in cells which contain DNA other than chromosomal DNA. These organellar

DNA carries genetic information and works independently or in collaboration with nuclear DNA (chromosomal DNA). Inheritance of characteristics from generation to generation by extrachromosomal / cytoplasmic/ organelle DNA is called cytoplasmic inheritance. There are a large number of examples which show the involvement of cytoplasmic DNA in controlling heredity traits of the organisms. Hence, they are also known as cytoplasmic heredity units or cytoplasmic genes.

These plasma genes are mostly received by the egg cytoplasm rather than the sperm cytoplasm. Hence, cytoplasmic inheritance is considered as a maternal inheritance phenomenon which influences the heredity characters. Though the cytoplasmic inheritance contributes to decide the characters of the offspring, reciprocal crosses do not result in the same phenotypes.

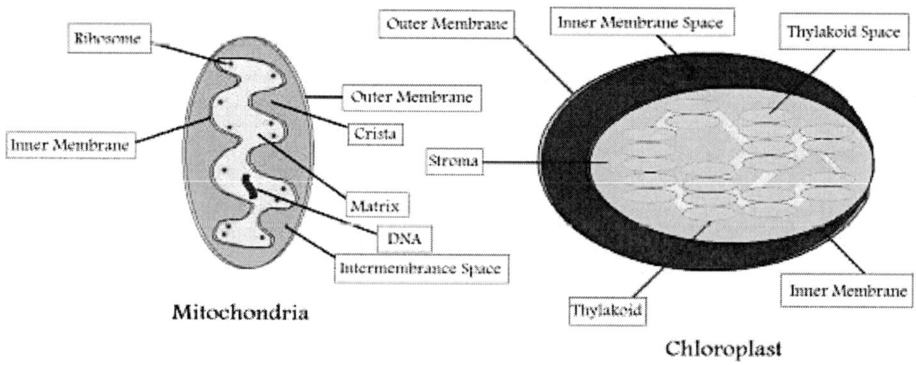

Figure: Mitochondria and Chloroplast

What is Genetic Maternal Effect?

Maternal effect is a situation which determines the phenotype of an offspring by the genotype of its mother, independent of the offspring genotype and environmental effect. In other words, the maternal effect is the casual influence of the maternal genotype on the phenotype of the offspring regardless of its genotype. It occurs due to the specific mRNA and proteins supplied by the mother to the zygote during embryo development. In many organisms, the embryo is initially inactive for the transcription. Hence, the supply of mRNA and proteins from the maternal side is important. The maternal effect does not arise due to heredity units. It arises totally due to these molecules received from the maternal supply. Due to these maternal effects, two offspring may sometimes differ phenotypically from each other though they possess the same genotype. One individual may resemble the mother parent.

The properties of cytoplasm are mainly governed by nuclear genes. Thus, the maternal effect depends on the nuclear genes.

Maternal effect is an important process in ecology and evolution. It contributes to population dynamics, phenotypic plasticity, niche construction, life-history evolution and natural selection.

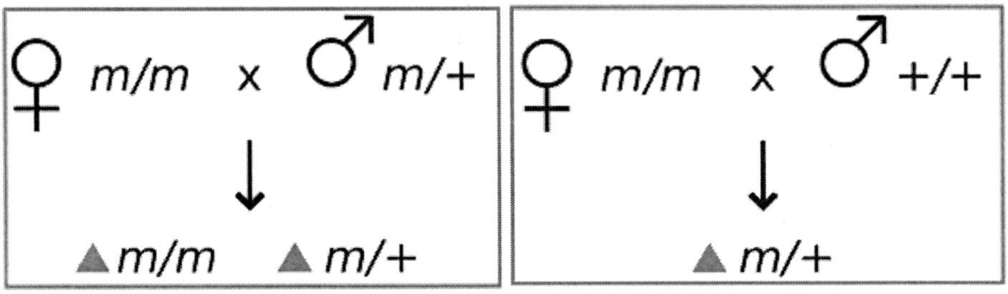

Figure: Genetic crosses involving a maternal effect recessive mutation

What is the difference between Cytoplasmic Inheritance and Genetic Maternal Effect?

Cytoplasmic Inheritance vs Genetic Maternal Effect	
Cytoplasmic inheritance is the inheritance of characteristics due to the genetic information stored in cytoplasmic DNA or organelle DNA.	Genetic maternal effect is the phenomenon where offspring traits are decided by maternal factors such as mRNA and proteins.
Occurrence	
Cytoplasmic inheritance is a result of actual genes received from mitochondria, chloroplasts or any infective particle like a virus.	Genetic maternal effect is a result of mRNA or proteins received from the mothers' egg.
Involvement of Organelles	
Cytoplasmic inheritance is involved in essential organelles like chloroplasts and mitochondria.	Genetic maternal effect is not involved in organelles.
Dependence on Nuclear Genes	
Cytoplasmic inheritance is not dependent on nuclear genes.	Genetic maternal effect may or may not dependent on nuclear genes.
Genetic Basis	
Cytoplasmic inheritance is due to actual genes.	Genetic maternal effect is due to gene products but not due to actual genes.

Summary – Cytoplasmic Inheritance vs Genetic Maternal Effect

Chromosomal DNA is considered as the sole genetic material of a cell. However several cellular organelles (mitochondria, chloroplasts) possess DNA which can influence the traits of the offspring. Some maternal products in the cytoplasm are also involved in deciding the traits of an offspring. Cytoplasmic inheritance and genetic maternal effect are two such situations. These two phenomena are caused due to the genes or factors inherited from the mothers' egg to the zygote. Maternal effect is a result of mRNA and proteins (gene products) received from the cytoplasm of the mothers' egg. Cytoplasmic inheritance is a result of genetic material in the mitochondria or chloroplasts or infective viruses. This is the main difference between cytoplasmic inheritance and genetic maternal effect. Offspring inherits the maternal traits regardless of its own genotype and the genes due to both these phenomena.

Meaning of Cytoplasmic Inheritance

Genes in the chromosomes have undoubtedly been proved to be responsible for the transmission of the various hereditary characters which in turn are located in the nucleus.

But in recent years there have been found some evidences which suggest that inheritance of characters occur by some self-perpetuating or replicating bodies like plastids and mitochondria which possess their own equipment for synthesizing for DNA and proteins present in the cytoplasm and not in the chromosomes present in the nucleus.

This particular type of mechanism in which cytoplasmic particles or inclusions take part in transmission of characters from generation to generation constitutes cytoplasmic inheritance.

The total self reproducing hereditary material of cytoplasm is termed as Plasmon like the genome (which refers to the total gene complement of an haploid set of chromosomes) of chromosomes and such units of cytoplasmic hereditary material are called cytoplasmic genes or plasma-genes or Plasmon's.

The cytoplasmic hereditary units are denoted by Greek letters- alpha, beta, gamma, sigma and so on, in contrast to the chromosomal genes which are denoted by Roman letters- a, b, c, d and so on. While considering various examples of cytoplasmic inheritance it has been assumed that the cytoplasm in the sperm cell is present in very minute quantity and mostly it is in larger quantity in the egg.

So, we could expect that plasma-genes mostly will transmit only through the egg cytoplasm rather than minute sperm. Thus, plasma-genes for various sizes and characters are likely to be supposed only in the egg cytoplasm.

Extra nuclear inheritance has two distinct features:

(i) It is more or less maternal inheritance i.e., only female contribute towards inheritance and therefore.

(ii) The result of reciprocal crosses are not the same. In these features, extra-nuclear inheritance contrasts sharply from nuclear inheritance.

Since the rediscovery of Mendel's laws in 1900, only a few geneticists believed that inheritance of characters take place through genes. Johannsen thought that genes determines only the superficial characters while fundamental characters of the organisms are transmitted through the cytoplasm.

Advances of our knowledge of heredity, however, left no doubt that genes in the chromosomes account for the specific transmission of most of the individual, racial and other characters.

Only in few cases, mostly in plants, evidences have been found that some traits or characters are transmitted through cytoplasm. Such traits are transmitted through the female line only. Each set of nuclear genes is sometimes designated as genome and all the hereditary materials transmitted through the cytoplasm is referred to as plasma-genes, cytoplasmic genes, cytogeneses, extra nuclear genes or extra chromosomal genes.

The sum total of the genes present in cytoplasm of a cell is known as Plasmon while all the genes present in a plastid are said as plastom. All available evidences indicate that plasma genes are found in the DNA present in mitochondria abbreviated as mt-DNA and in chloroplast shortened as Cp- DNA.

Evidence for cytoplasmic inheritance was first presented by Correns (1908) in Mirabilis jalapa and by Baur in Pelargonium zonale. Both of these geneticists studied the inheritance of variegation in the respective plant species.

Correns in his experiment found that only the character of female parent was observed in to the progeny. But the examples of such a role were so few that Morgan in 1926 concluded that "the cytoplasm may be ignored genetically."

Examples of Cytoplasmic Inheritance:

Maternal Inheritance:

In Ephestia moth:

A clear case in which individual differences are known to depend on the extra nuclear transmission has been studied in the colour of the larval skin and eye of the moth Ephestia.

The colour depends upon the presence of pigments. The synthesis of pigment is controlled by a hormone like substance known as Kynurenine. The hormone like substance is controlled by allele 'A' and the absence of the substance is due to the presence of allele 'a' in a homozygous condition.

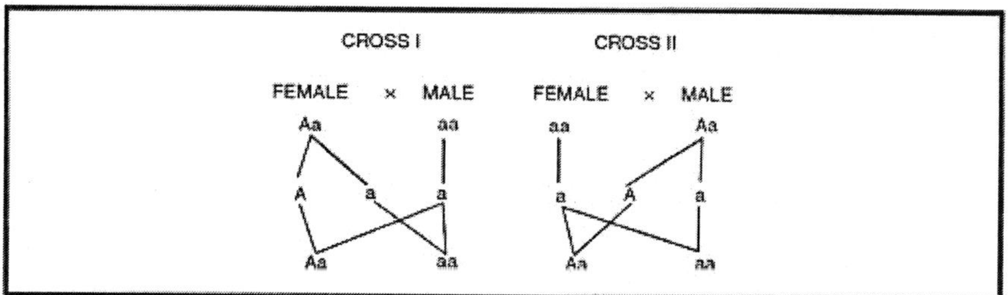

Fig. 109. Cross I—both 'Aa' and 'aa' pigmented. Cross II—'Aa' pigmented and 'aa' non-pigmented.

In the progeny of cross I and II, individuals of the constitution aa, should not show any pigment but the situation is some what different. The progeny of cross I with the constitution 'a' develop pigments in the skin and eyes in the larval stage; while those of cross II shows no pigment in the larval stage.

So, it became very clear that the presence or absence of pigment seems to depend upon the type of mother and on the cytoplasm, which offspring got through the ovum. Having got it through the cytoplasm, it may be cited as an example of cytoplasmic inheritance.

The real explanation is different. Allele 'A' elaborates the hormone like substance, kynurenine and this finds place in both type of ova 'A' and 'a' at the time of oogenesis. The small amount in the ovum 'a' is responsible for the development of pigments in the larva with the constitution aa.

But the pigments thus developed is only temporary and is lost in the adult. Since this is not permanent and is not transferred to subsequent generations. Some authors consider that this is not example of cytoplasmic inheritance at all.

In snail Limnaea:

An example of inheritance determined by female (mother) eggs is described in water snail Limnaea by Boycott Diver and Garstang on its coiling. The dextral or right coiling is represented by dominant (DD) while its recessive is sinistral or left coiling (dd).

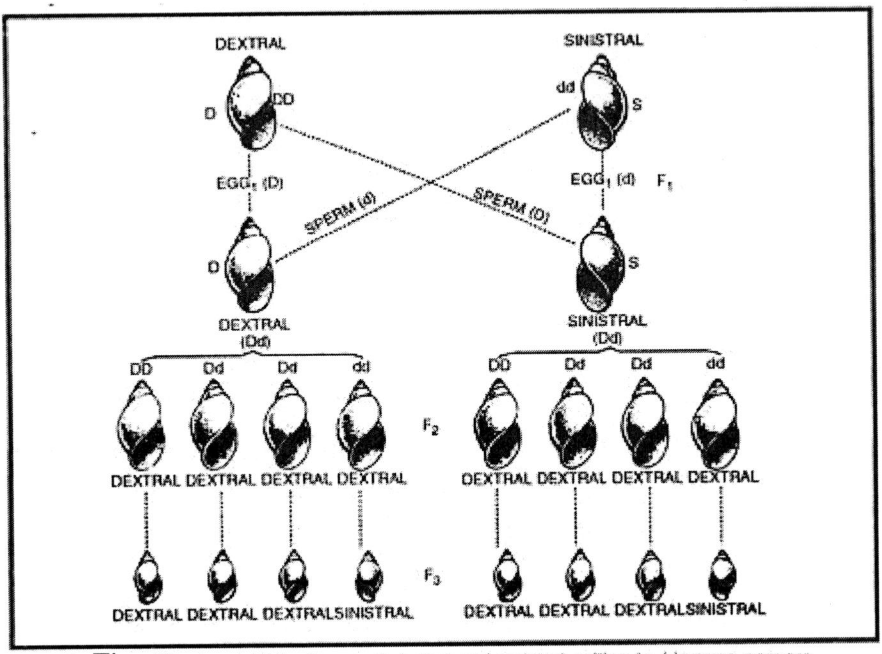

Fig. The normal inheritance of dextral and sinistral coiling in *Limnaea peregra*. D=Dextral phenotype, S=Sinistral phenotype, D = gene for dextrality, d = gene for sinistrality. The species is bisexual (⚥) and may reproduce either by crossing or by self fertilization

Now the direction in which the developing egg coils is not determined by its own genes but by the genes of its mother. Suppose mother is left turning (sinistral) designated as dd, its eggs will also be left turning coils while father is right turning (dextral) DD.

When a cross occurs between sinistral female and dextral male, the eggs will develop in to sinistral females like the mother even if fertilized by sperms of dextral males. This maternal influence is probably due to presence of some cytoplasmic bodies in the mother's egg cell. This phenomenon is called maternal determination.

Kappa Particles in Paramecium:

Some varieties of Paramecia produce a toxic or lethal substance called paramecin that kills other varieties. In a series of investigations,

T.M. Sonneborn and others found that this characteristic involves a kind of extra-nuclear inheritance. Parmecia that produce the toxic or lethal material are called killers and those individuals that are killed are known as sensitives or normal.

The killer race contains certain bodies called as kappa bodies which are directly responsible for the production of killing substance. The normal or sensitives lack these particles. The Kappa particles pass from one generation of Paramecia to the next in the process of cell division.

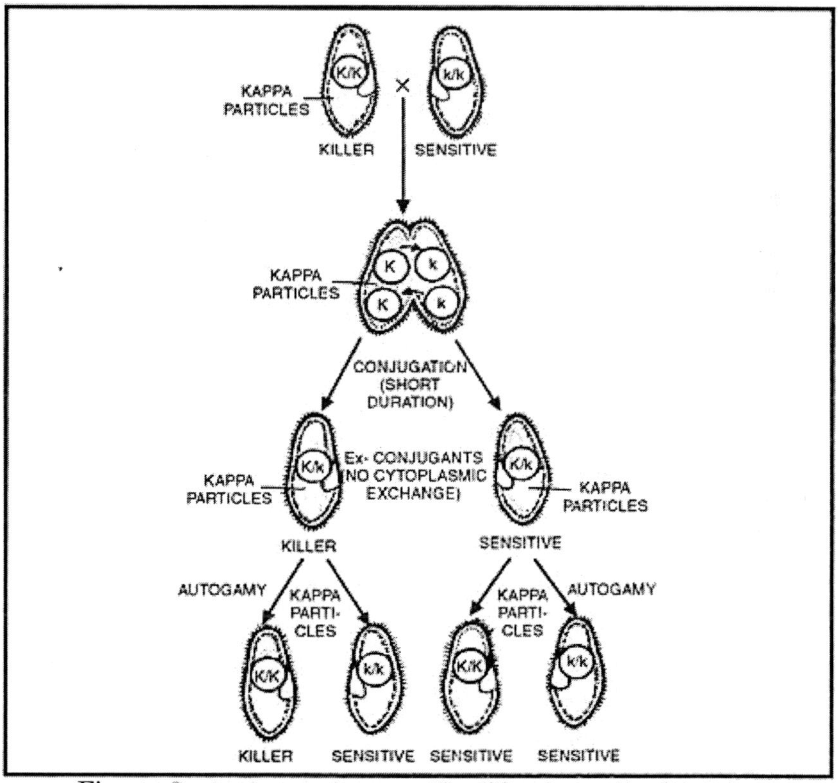

Fig. (a) Results of a cross between a killer (KK) and a sensitive (kk) strains of *Paramecium*, when no cytoplasmic exchange is allowed.

The continued reproduction or existence of kappa in a Paramecium depends on the presence of the dominant gene 'K' in the nucleus and the sensitives have the recessive allele 'k'.

These genes are strictly transmitted as nuclear genes and kappa particles are transmitted through the cytoplasm. When once kappa is lost, it can not be regained, unless more kappa is introduced from another cell i.e., gene 'K' can only maintain the kappa particles and can not initiate its production.

When killers KK conjugate with sensitives kk, the ex-conjugants are Kk and should be killers, but if the conjugation is brief and little or no cytoplasm is exchanged, then a killer clone and a sensitive clone are produced. A more prolonged conjugation results in exchange of cytoplasm and all descendent clones are killers and have kappa particles.

So it is clear that these particles are clearly vehicles of cytoplasmic transmission and it has been shown that maintenance of these particles and the killer phenotype depends upon the presence of gene K in the nucleus. Thus a 'Kk' individual may be either a killer or a sensitive depending on whether or not it has received kappa particles.

Certain killer clones 'KK' or 'Kk' can be converted in to sensitives by making them undergo very rapid fissions (amitosis cell division as in bacteria or prokaryotes) at high temperature, during which the reproduction of kappa fails to keep distance or pace with the cell divisions.

As a result of this, the number of kappa per Paramecium becomes fewer and fewer in the course of the cell division and eventually some paramecia are left without any kappa particles and become sensitives. Thereafter, the gene 'K' can not initiate the production of the particles but can again maintain them if they are introduced during conjugation.

The exact nature of kappa has not been known. Some biologists consider that kappa is an symbiont organism and others think that it might be a gene present in the cytoplasm. So at present it is impossible to decide between the two theories.

In any case, kappa is capable of specific self duplication; it is found in the cytoplasm of the cell and it is normally transmitted from generation to generation. It has also been shown that kappa particles contain DNA.

Until the true nature of kappa is definitely settled, kappa can not be considered as an evidence for cytoplasmic genes. E. Altenburg suggests that kappa was derived from Zoochlorellae and in the course of evolution had lost their chlorophyll and assumed additional properties like killing etc.

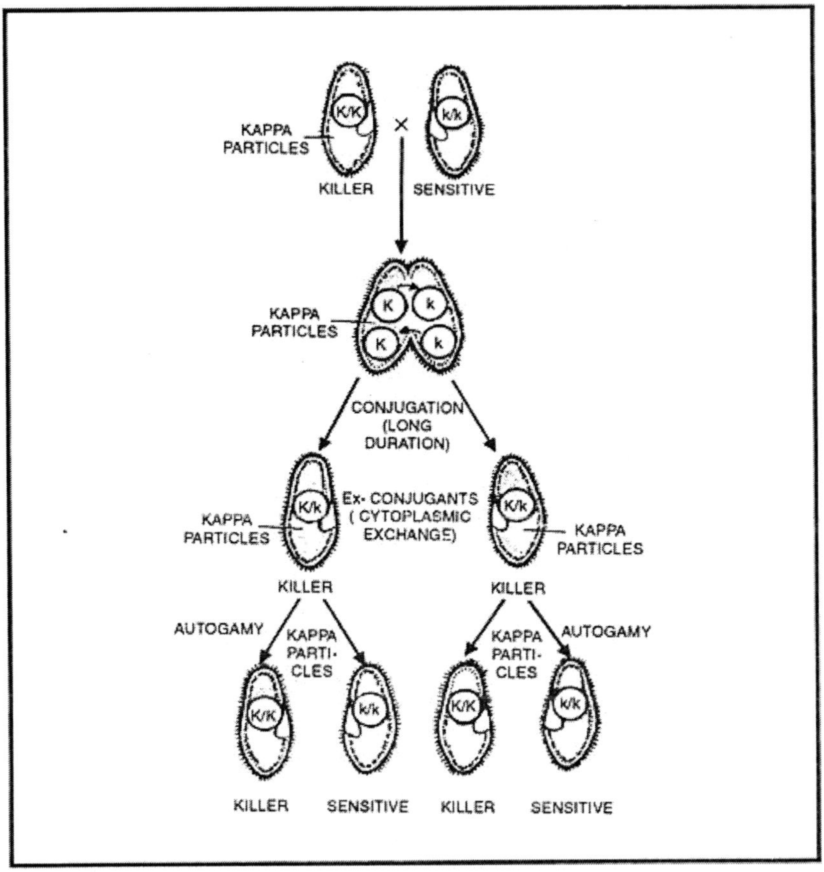

Fig. (b) Results of a cross between a killer (kk) and a sensitive (kk) strain of Paramecium, when cytoplasmic exchange is allowed

CO_2 Sensitivity in Drosophila:

This is another example to show the cytoplasmic inheritance in animals. Normally Drosophila can be anaesthetized by CO_2 and they can tolerate higher concentration of the gas without injury.

But Heritier and Teissier discovered a race of Drosophila which is much more sensitive to CO_2 than are the normals and they become anaesthetized on exposure to CO_2 in much less time than do the normals or sometimes they are even killed.

This feature is called CO_2 sensitivity, and is normally transmitted through the egg. Since the type of the offspring entirely depends upon the types of female parents in the cross; it has been cited as an example of cytoplasmic inheritance. Heritier was also able to produce sensitivity in to normal fly either by the grafting the organs of sensitive fly in to normal fly or by injecting extracts of sensitive fly in to normal.

But recent observations have led to the conclusion that CO_2 sensitivity is due to a virus. All the cells or almost of the cells of body, both somatic and germinal contain the virus, but the eggs with their larger amount of cytoplasm contain more virus particles than do the sperm cells.

Therefore, the eggs, transmit the infection much more efficiently than do the sperm. Since CO_2 sensitivity depends upon a virus, this can not be considered as a case of cytoplasmic inheritance.

Plastid Inheritance:

The cytoplasm of plants contain minute bodies called plastids. Some of them contain chlorophyll or green pigment produced by chloroplastids. Others are colourless and produced by leucoplastids and contain starch. Plastids resemble genes in two respects (i) they are capable of multiplying by division and (ii) they are capable of mutation.

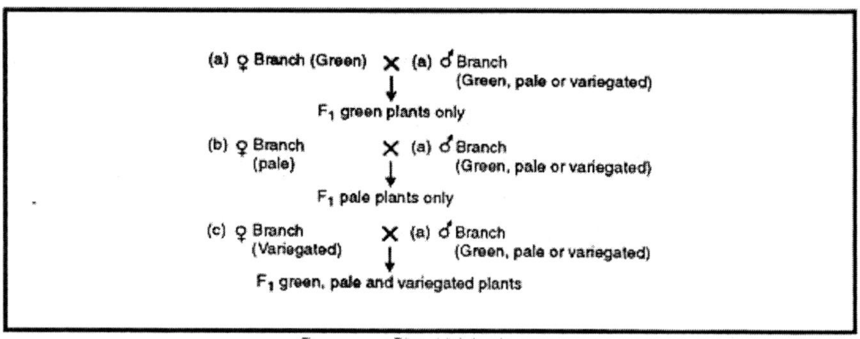

Fig. Plastid inheritance

The best known cases of continuous transmission of stable elements through the cytoplasm come from the plastid inheritance of plants. Correns studied the albomaculatus type of leaf variegation in the four O' clock plant, Mirabilis jalapa in which the normal green tissue irregularly spotted with patches of pale green or white.

Flowers on wholly green branches produce seeds which grow in to normal plants. Flower of variegated plants yield offspring of 3 kinds viz., green, pale and variegated in various proportions. Flowers from wholly white give progeny without chlorophyll. In each case the pollen has no influence on the offspring. Inheritance is wholly maternal.

Progeny of a variegated four o'clock plant.

Pollen from branch of type	Pollinated flowers on branch of type	Progeny grown from seed
Pale or white	Pale, green and variegated	Pale, green, Pale-green & variegated
Green	Pale, green and variegated	Pale, green, Pale-green & variegated
Variegated	Pale, green and variegated	Pale, green, Pale-green & variegated

The colour of the offspring (Pale, green, pale-green and variegated) is determined by the egg and what the egg produces, depends upon the characters of its plastids. Since plastids are found in the cytoplasm, we can cite this as an example of cytoplasmic inheritance. Moreover, plastids are plasma genes since they have the capacity of self duplication and mutation like the genes.

Male Sterility in Plants:

Besides numerous examples where the plastids have been found as definite means of extra nuclear transmission of characters, there are also many other cases where cytoplasm influences the offspring but what exactly the cytoplasm has property is still not understood.

Certain types of male sterility have been found in corn, onion etc. This feature seems to depend upon a considerable degree on cytoplasmic control, but no definite cytoplasmic elements like plastids or any other elements have been found to be responsible for this character.

This character finds expression in the total abortion of all pollen and never the ovules. The exact mode of inheritance of this character seems to be entirely through the mother.

A classical example of maternal inheritance of male sterility was

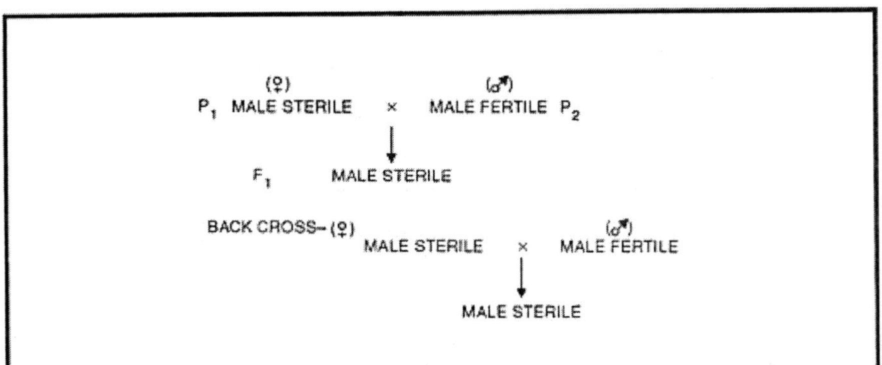

Fig. Maternal inheritance of cytoplasmic male sterility in plants

discovered by MM Rhoades in Maize. He successfully replaced all the chromosomes of the male sterile plant with those of the normal and still found male sterility persisting in plants and this indicates that male sterility as a character or trait is resident in the cytoplasm and hence this may be cited as an example of cytoplasmic inheritance.

Merogonic Hybrids:

Merogonic hybrids are those where the egg cytoplasm of one species is fertilized by the sperm nucleus of another species. Theodor Boveri removed the nucleus from the egg of sea-urchin of one species and artificially fertilized them by sperms of a species with different characters.

The merogonic hybrids seemed to resemble the species that furnished the sperm. This demonstrates that the nucleus controlled the development even when surrounded by cytoplasm from a very different source.

The work of Hadorn, however, rules out this monopoly of the nucleus over all traits in the offspring. Two species of Triton differ clearly in the character of the epidermis. Hadorn fertilized the egg of Triton palmatus with the sperm of Triton cristatus and succeeded in removing the T. palmatus nucleus before nuclear fusion. Development proceeds up to the blastula stage and stops.

If the presumptive epidermis from the blastula is grafted on to the normally developing larva of Triton alpestris, it maintains its identity and develops into the adult stage. Here it resembles the epidermis of the parent. T. palmatus contributed the cytoplasm where as T. cristatus contributed the nucleus. This shows that at-least in this trait e.g., the nature of the epidermis is under cytoplasmic control and to this extent the nucleus does not hold monopoly.

Milk Factor in Mice:

In outcrossing mice from a long inbred strain in which nearly all females developed a particular type of mammary cancer. J.J Bittner (1940) found that cancer susceptibility was maternally transmitted.

Moreover, when baby mice from lines showing respectively high and low incidence of cancer were taken from their mothers at birth, and allowed to nurse on foster mothers of the opposite type, it was shown that the substance that induces cancer susceptibility is transmitted through mother's milk i.e., mother mice from lines showing a high incidence of the disease transmit through their milk a substance that later causes mammary cancer to develop in mice that have nursed from them.

These results established a maternal inheritance dependent on a factor carried in the milk. More recent studies have shown that the factor is transmitted also by body fluids other than milk including saliva and semen. Recent evidences suggest that this disease is probably due to virus. So this offers no support for the theory that the cytoplasm contains genes.

Predetermination and Dauermodifications:

Mostly maternally transmitted characters are due to cytoplasmic inheritance. However, there are certain cases where the maternally transmitted cases are not governed by cytoplasmic inheritance. The egg cell which develops within mother's body, its cytoplasm is well affected by the mother's gene as well as by the substances produced within mother's body itself.

All of these affect some of the characters of embryo. This effect of the maternal genes on the cytoplasm of egg cell is called predetermination and the characters affected are said to be predetermined. Example of this type is the direction of coiling in snails and development of cancer in the breast of female mice at maturity stage. The gene for this case is transmitted to the young ones through the mothers milk.

Occasionally phenotypic changes are produced as a result of the environmental factors or chemicals and are transmitted through the cytoplasm of the egg cell. The changes so far produced insist for a few generations but gradually comes to declinity and finally get disappeared. These are called dauermodifications. These are never gene controlled characters and disobey Mendelian traits.

Cytoplasmic inheritance in protozoa

Cytoplasmic Inheritance Overview

Cytoplasmic inheritance meaning is the inheritance of organelle DNA in the offspring from the parent's DNA. Cytoplasmic inheritance is different from nuclear genetics because it does not follow the laws of gene inheritance. Half of the genes will come from male and female parents according to gene inheritance. However, depending on the species, DNA found in the organelles can come exclusively from males, females, or an unequal mixture of both.

Cytoplasmic inheritance is governed by genes found in chloroplasts and mitochondria.

Cytoplasmic Inheritance Types

They are of two primary types:

1. **Plastid inheritance**
   - Chloroplasts are an essential type of plastids
   - Applicable only on plants, as plastids are only found in plants

   Some examples:

   Mirabilis jalapa:

   - The *Mirabilis jalapa* plant has three different types of leaves-green, white and variegated leaves
   - In the case of *Mirabilis jalapa*, inheritance is administered by chloroplasts
   - In *Mirabilis jalapa*, the green leaf has normal chloroplasts. The white leaf has mutant chloroplasts, whereas the variegated leaf has a combination of mutant and normal chloroplasts

   Oenothera:

   - In Oenothera, the whole chromosome set of pollen parent or pistillate of an individual is transferred to the gametes
   - Gametes contain either all chromosomes of a male or a female parent
   - Such inheritance of chromosomes is observed because of the series of reciprocal translocations

   Iojap in Maize:

   - Plastids govern the soap phenotype
   - In the case of Iojap in maize, normal plastids are found in green leaves. Mutant plastids are found in white leaves, while a mixture of normal and mutant plastids is found in the striped leaves
   - There are three different categories of plastids for a cross between the green male and the Iojap female- white, green, or both

2. **Mitochondrial inheritance**
   - Mitochondrial inheritance is common for both plants and animals

Some examples are:

Cytoplasmic male sterility in plants:

- There are three types of male sterility in crop plants: cytoplasmic, genetic, and cytoplasmic
- In maize, mitochondrial DNA determines cytoplasmic sterility
- Cytoplasmic sterility is detected in many other crop plants such as cotton, sorghum, pearl millet, etc.

Pokiness in Neurospora:

- Neurospora is a bread mould. It has two strains, the first is wild, and the second is poky
- The wild strain part of Neurospora has normal growth, whereas the mutant or poky part has very slow growth
- Suggesting the existence of cytoplasmic inheritance. It is due to the only distinction between the reciprocal crosses
- Females contribute the most to the cytoplasm in Neurospora

Petite in yeast:

- Petite yeast, also known as little yeast, is Ascomycetes
- Petite is very small in size
- Petite is a mutant, i.e., different from the normal conditions
- Cytoplasmic factors control this mutation

Characteristics

1. Reciprocal Differences: Cytoplasmic inheritance governs the characters with notable distinctions in reciprocal crosses in F.
2. Maternal Effects: Distinct maternal effects are observed due to more contribution of cytoplasm to the zygote by a female parent than a male parent.
3. Mappability: Chloroplast genes and mitochondrial genes have been mapped, which were very difficult to map. Chloroplast genes in Chlamydomonas and maize and mitochondrial genes in humans and yeast are some of the examples.
4. Non-Mendelian Segregation: A typical segregation pattern not followed in cytoplasmic inheritance exhibits in the Mendelian inheritance.

5. Somatic Segregation: The characters, which are controlled by cytoplasmic genes, display segregation in somatic tissues. For example, segregation in somatic tissues like leaf variegation.
6. Governed by plasma genes: Chloroplast or mitochondria DNA governs or manages Cytoplasmic inheritance.

Significance

- Cytoplasmic inheritance helps determine the significance of cytoplasmic organelles in the passing on of characteristics in several organisms.
- Cytoplasmic inheritance helped map the chloroplast genome and mitochondrial genome in many species, such as humans, yeasts, maize, etc.
- Cytoplasmic male sterility has developed in crops like maize, cotton, pearl millet, etc. It has promoted the production of hybrid seeds.
- The role of mitochondria is gaining importance in heterosis, as heterosis results in vigorous growth of plants, which ultimately provides a higher yield of crops.
- The mutation leads to the generation of new variants of characteristics. In cytoplasmic inheritance, these mutations are observed in chloroplast and mitochondrial DNA.

The important features of cytoplasmic DNA in comparison to nuclear DNA are briefly described below:

1. The cytoplasmic DNA is of two types, viz., chloroplast DNA (cp-DNA) and mitochondrial DNA (mt-DNA).
2. The cytoplasmic DNA is usually circular molecule except in ciliate protozoa where it is linear molecule. Nuclear DNA on the other hand is usually linear in eukaryotes and circular in prokaryotes.
3. The synthesis of chloroplast DNA and mitochondrial DNA continues throughout the cell cycle, whereas synthesis of nuclear DNA occurs only during interphase (S) of cell cycle.
4. Cytoplasmic DNA replicates both in chloroplasts and mitochondria in a semiconservative fashion, whereas nuclear DNA replicates in the chromosomes in semiconservative manner.

5. Both cytoplasmic and nuclear DNAs are liable to mutation. Mutation of cytoplasmic DNA leads to change in the behaviour of plasma genes and mutation of nuclear DNA leads to alteration in nuclear genes.
6. Cytoplasmic DNA is capable of coding for RNA and protein synthesis like nuclear DNA.
7. In vitro synthesis of cytoplasmic DNA is possible in the respective organelle. In vitro synthesis of nuclear DNA is also possible.
8. The synthesis of cytoplasmic DNA is blocked when treated with ethidium bromide, acriflavin and actinomycin-D. Similar is the case with nuclear DNA. Thus, cytoplasmic DNA differs from nuclear DNA in four main aspects.

Inheritance of Preformed Structures:

In some lower group of animals like ciliated protozoa there are some characteristics which show cytoplasmic inheritance. These traits are not controlled by any plasma gene. In Paramoecium, the cytoplasm is differentiated into cortex or ectoplasm and medulla or endoplasm.

The mouth and the contractile vacuole are the prominent preformed traits that are present in the cortical region of Paramoecium. These preformed structures can be transmit-ted independent of the transmission of nuclear genes and plasma genes.

For normal sexual reproduction two individ-uals of Paramoecium called conjugants come close together, exchange their nuclear material, and then separate as ex-conjugants with zygote nucleus. In some rare cases, conjugants do not separate and remain as doublet animal with two sets of mouth and contractile vacuole structures.

When this doublet animal repro-duces asexually by binary fission, they also give rise to doublet ex-conjugants. When doublets are mated with normal singlet's, the progeny of doublets ex-conjugants are doublets and the progeny of singlet's are singlet's. This type of inheritance is also found when the animals reproduces through autogamy.

In these crosses it is noted that the nuclear genes of doublets and singlet's are inherited in normal Mendelian fashion. But the mode of duplicated mouth and contractile vacuole is independent of the mode of inheritance of nuclear genes as well as cytoplasmic hereditary factors.

Furthermore, the transplantation ex-periments reveal that such cortical structure (mouth, contractile vacuole) are autonomous and are maintained by cell division.

Some born has suggested that different parts of the cortex might serve as sites for the specific absorption and orientation of molecules derived from the milieu and genetic action. Besides this, preformed cortical structures could act by determining where , some gene products go in the cell, how these combine and orient and what they do.

Extra-Nuclear Inheritance in Bacteria:

In many cases, extra-nuclear inheritance has also been reported in bacteria. In addition to main chromosome, bacterial cell has got a unique extra-nuclear genetic system in their cytoplasm. Such extra-nuclear genetic system plays an important role for cytoplasmic basis of inheritance.

Organelles genetics inheritance – chloroplast mitochondria

Chloroplast inheritance: Early experiments

At the turn of the 20th century, Carl Correns, a German botanist, did a series of genetic experiments using four o'clock plants (*Mirabilis jalapa*). We now know that his work demonstrated how chloroplast DNA is passed on from cell to cell and from parent to offspring—though Correns himself didn't know it at the time55start superscript, 5, end superscript!

Correns' experiments

The *Mirabilis* plants that Correns worked with came in three types: pure green, pure white, or variegated (mottled green and white). Green and white branches could appear on variegated plants, but variegated branches did not appear on green or white plants66start superscript, 6, end superscript.

Correns was curious about this coloration trait, and he carried out a number of crosses between plants of different colors. He found that66start superscript, 6, end superscript:

- The color of the egg cell-donating branch (female parent) determined the color of the offspring.
- Female parent branches that were pure green or pure white produced only pure green or pure white offspring, respectively.
- Female parent branches that were variegated could produce all three types of offspring, but not in any predictable ratios.

Correns speculated that some factor in the cytoplasm of the egg cell must determine the color of the offspring. It was actually a different

German botanist, Erwin Baur, who suggested that the chloroplasts in the cytoplasm might carry hereditary factors (genes).

Baur thought that, in variegated plants, some of the chloroplasts must have mutations that made them unable to turn green (produce pigment). Today, we know that this hypothesis was exactly right!

Structure of Organelle Genomes

One basic concept of biology suggests that eukaryotic organelles arose after a primitive eukaryotic cell engulfed (swallowed, invaginated) a prokaryotic cell. **This is called the endosymbiotic theory.** After this invagination event, the concept continues, duplicate functions already possessed by the eukaryotic cell were eliminated, and only those prokaryotic functions that were advantageous to the eukaryotic cell were maintained. In particular, energy transduction functions, such as ATP and NADPH production, were maintained.

The two organelles that arose were the mitochondria, found in all eukoryotic cells, and chloroplasts which are found in plants and algae. Some of the genetic information of the prokaryotic cell was transferred to the nucleus of the eukaryotic cell. Chloroplast and mitochondrial sequences have been found in the nucleus of plant cells. Furthermore, chloroplast sequences have been found in the mitochondrial of plant cells. This is clear evidence that genetic control of certain biochemical functions was relinquished (or taken from) the progenitor cells. The following diagram shows the flow of genetic information from organelles in a plant cell.

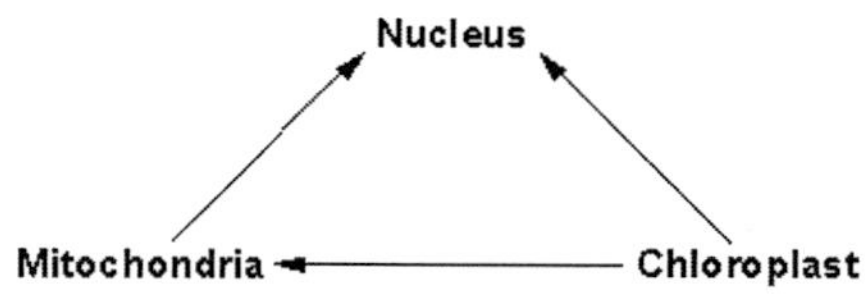

It is important to note that all the basic functions of the Central Dogma of Molecular Genetics are found in organelles. This functions include DNA replication, RNA transcription and protein translation. Thus, certain gene products will result from the expression of the organelle DNA. The number of protein products from mitochondrial transcription is limited. One set of genes that are expressed are the rRNAs and tRNAs that are required for translation. Five known gene products are produced from the mammalian mitochondrial genome. These include subunits I, II and III for cytochrome oxidase, the apoprotein for cytochrome b and subunit 6 of the mitochondrial ATPase.

In addition to these gene products, six ORFs have been identified. (ORF represents Open Reading Frame.) These are sequences that have transcription start and stop signals that bracket sequences that could produce a protein product, but the actual function of the product is not known. Finally, as we saw above the expression of organelle genes has a unique effect on the inheritance of certain traits.

This transfer of genetic information during the evolution of eukaryotic cells has also required the development of cooperative gene expression systems between the organelle and the nucleus. Cytochrome oxidase is one of the mitochondrial enzymes involved in ATP generation. As stated above subunits I, II, and III are encoded in the mitochondria. The remainder of the subunits of this seven subunit protein are encoded in the nucleus. The best studied example of coupled nuclear DNA/ chloroplast DNA gene expression is for the protein RUBISCO (ribulose bisphoshpate carboxylase oxygenase). This is the enzyme that begins the fixation of atmospheric CO2 into sugar molecules. This enzyme has 16 subunits, eight large subunits and 8 small subunits. The large subunits are encoded by the chloroplast DNA whereas, the small subunits are encoded by nuclear DNA. Thus, for those proteins that are encoded by genes in two different cellular locations, gene expression has to be coordinated between the two locations for proper protein functioning.

Chloroplast Genomes

The chloroplast genomes of plants exhibit a far greater conversation of structure than plant mitochondrial genomes. These genomes are circular and the size of higher plant chloroplast DNAs are either *150 kb (for example, spinach) or *120 (for example pea). The difference in size can be accounted for by a deletion from the larger genome to generate the smaller. The gene order among all higher plant chloroplast genomes is essentially conserved.

As you would predict, many of chloroplast DNA genes encode proteins that are involved in photosynthesis. In total, the genome appears to encode for a complete set of rRNA and tRNA genes and *45 protein products. There are some differences between species, but these differences primarily are between higher plants and algae, which also contain chloroplast DNA.

RFLP proof that chloroplast DNA can be maternally inherited

As has been stated previously, to demonstrate that a trait is maternally inherited specific crosses need to be made to generate the required offspring. A number of crosses where made between cultivated tomato (L. esculentum) and a number of wild species. As we discussed previously, chloroplast DNA is a circular molecule *150 kb in size. Digestion of this molecule produces relatively few fragments. Within the plant cell, chloroplast DNA represents a significant portion (*15%) of the DNA. This is a result of the large number of chloroplasts per cell (*50) and the large number of chloroplast DNA molecules per chloroplast (*150).

Chloroplast DNA was obtained from F1 plants of a number of crosses in which L. esculentum was the female in the cross. As can be seen below, in each case the F1 restriction fragment pattern was identical to L. esculentum (sample 8). This is conclusive evidence that chloroplast DNA is inherited in a maternal manner.

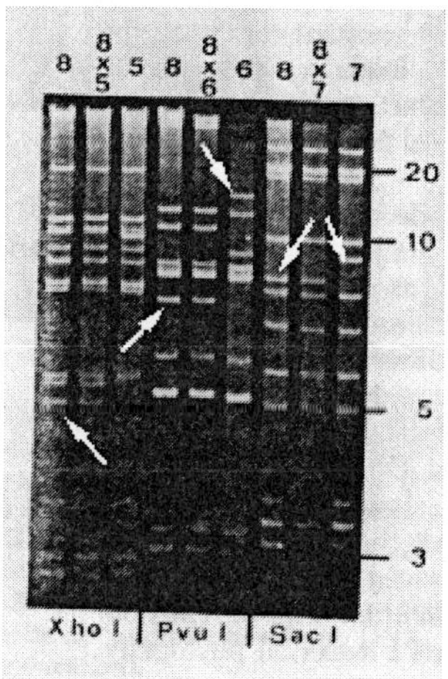

Mitochondrial Genomes

All of the eukaryotic species that have been analyzed to date have DNA in the mitochondria. But the size of this mitochondrial genome varies.

Size of Mitochondrial G enomes

Species	Size (kb)
Human	16
Drosophila	18
Yeast	75
Turnip	218
Corn	570
Muskmelon	2000

All of the mitochondrial genomes are circular. Even though there is a large size discrepancy between different species, especially between plants and animals, the number of genes that are expressed in each species is nearly the same, about 20. Thus the extra DNA that is found in the larger genomes does not appear to be required. Studies have shown that much of this DNA is in fact repeated sequences.

RFLP proof that mitochondrial DNA can be maternally inherited

Experiments designed to show that mitochondrial DNA was maternally inherited were analogous to those which demonstrated that chloroplast DNA was maternally inherited. Mouse is the experimental organism used to demonstrate this principle here. Reciprocal crosses were made between the species Mus domesticus, common mouse, and the wild species Mus spretus. F1 progeny where then backcrossed, with one of the two species serving as females. The restriction enzyme HincII is diagnostic for the mitochondrial DNA of the two species. M. domesticus mt DNA contains five restriction sites for the enzyme that generates four fragments whereas M. spretus mt DNA contains eight restriction sites which generate seven fragments. Only two of the fragments are the same size and presumably contain the same sequence information.

The results of the backcross experiments. As you can see the female contributes the mitochondrial genome. These experiments have been repeated in other species, such as human, and the female has been shown to contribute the mitochondrial genome.

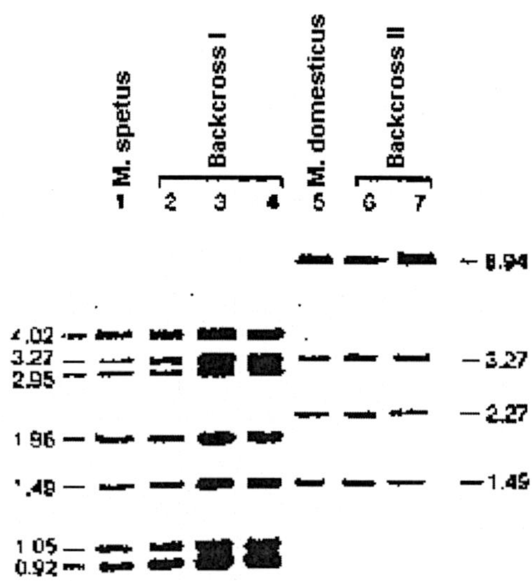

Chapter 10

Genetics of Breeding and Heterosis

An Introduction to Heterosis

Plant breeding is the application of genetic principles in developing new plant varieties, known as cultivar development, crop improvement, and seed improvement. Heterosis in plant breeding is described as the superiority of an F_1 hybrid over both parents in terms of yield or other characteristics. Heterosis contributes to increased vigour, size, growth rate, yield, or other attributes. However, in exceptional cases, the hybrid may be inferior to the weaker parent. The methods of estimation of heterosis and the genetic basis of heterosis are described here.

Heterosis Definition

Heterosis refers to the superiority of F, hybrids over their parents in one or more characteristics. The word hybrid vigour is a synonym for heterosis. George Harrison Shull coined the term heterosis in 1914.

Some features of heterosis are described below.

- **Superiority Over Parents:** Heterosis results in superiority over its parents in adaptability, yield, quality, disease resistance, maturity, and general vigour. Positive heterosis is often seen as desirable. However, in some circumstances, negative heterosis is preferable. Negative heterosis for plant height, maturity time, and hazardous chemicals, for example, is beneficial in many circumstances since it demonstrates superiority over the parents. In most agricultural plants, heterosis of 40% or more over the superior parent is regarded as substantial from a practical standpoint.
- **Confined to F_1:** Heterosis is restricted to the F_1, resulting in the production of a cross. As a result of segregation and recombination, it diminishes and vanishes in F_1 and later generations of a cross. As a result, heterosis is linked to the F_1 generation.

- **Genetic Control:** Nuclear genes regulate the expression of heterosis. In certain cases, heterosis is caused by the interaction of nuclear genes and cytoplasm.
- **Reproducible:** Once recognised, heterosis may be easily reproduced in a specific environment. The expression of heterosis, on the other hand, is more evident in the region of hybrid adaptability.
- **Relationship with SCA:** Heterosis shows a positive relationship with specific combining ability (SCA) variation. The SCA is a measure of dominance variance, and having a high degree of dominance variance is required to carry out a heterosis breeding program.
- **Heterozygosity Effect:** The degree of heterosis is related to heterozygosity since dominance variance is related to heterozygosity. The dominance effects should be most significant in cross-pollinated species and least significant in self-pollinated species. As a result, heterosis occurs more frequently in cross-pollinated crops than in self-pollinated crops.
- **Masks Recessive Genes:** When there is heterosis, the beneficial influence of dominant genes masks harmful recessive genes. As a result, recessive mutant genes are concealed in heterozygous individuals.
- **Low Frequency:** The frequency of good heterotic pairings is quite low. Only a few good heterotic pairings are discovered after screening thousands of F_1 crosses. All of the F_1 crosses lack desired heterosis.

Genetic Basis of Heterosis

To explain the mechanism of heterosis, two significant theories have been suggested. The first is the dominance theory, while the second is the overdominance hypothesis. Epistasis is also probably related to heterosis. As a result, there are three potential genetic origins of heterosis, which are:

- Dominance
- Overdominance
- Epistasis

Dominance Hypothesis

Davenport (1908), Bruce (1910), and Keeble and Pellew (1910) proposed this hypothesis. This is the most commonly accepted explanation for heterosis. According to this theory, heterosis is caused by the superiority of dominant alleles when recessive alleles are harmful. The hybrid shows heterosis because the deleterious recessive genes of one parent are concealed by the dominant genes of the other parent. Both parents have different dominant genes.

Assume one parent's genetic make-up is AABBccdd and the other's is aabbCCDD. A hybrid of these two parents will have four dominant genes, giving it superiority over both parents having two dominant genes. Thus, heterosis is proportional to the number of dominant genes contributed by each parent.

Dominance Hypothesis

Overdominance Hypothesis

Shull and East separately presented this hypothesis in 1908. This hypothesis is known as stimulation of heterozygosis, cumulative action of divergent alleles, single-gene heterosis, super-dominance, and overdominance. Even though Shull and East proposed this hypothesis in 1908, Hull used the word overdominance in 1945 when working on maize. According to this theory, heterosis is caused by the heterozygote's superiority over both of its homozygous parents. Thus, heterosis is proportional to heterozygosity.

The superiority of the heterozygote over both homozygotes may result from:

- The production of a superior hybrid substance in the heterozygote that is entirely different from either of the homozygous.
- Greater buffering capacity in the heterozygote due to cumulative action of divergent alleles or stimulation of divergent alleles. East explained this theory in 1936, suggesting a set of alleles a1, a2, a3, and a4 with steadily increasing divergence in function. As a result, a combination of more divergent alleles will have more heterosis than a combination of less divergent alleles. Combinations of a1a4, for example, demonstrate more heterosis than combinations of a1a2, a2a3, and a3a4. Overdominance has been reported in barley.

Epistasis

The interaction of alleles from two or more distinct loci is referred to as epistasis. It is sometimes referred to as nonallelic interaction. Non-allelic interactions are classified into three types: additive x additive, dominance x dominance, and additive x dominance. It is widely documented that the presence and size of non-interaction have a positive relationship with the incidence and magnitude of heterosis. Epistasis, especially dominance effects (dominance x dominance), may lead to heterosis. Cotton and maize have both shown this (Moll and Stuber 1974). Various biometrical models can detect or estimate epistasis.

Consequences depression - inbreeding coefficient(f)

Decreasing fitness with increasing inbreeding is a widespread phenomenon , which occurs because mating between relatives increases homozygosity at loci carrying rare recessive deleterious alleles or exhibiting overdominance . The magnitude of inbreeding depression is relevant to many disciplines within biology, including the conservation of small, isolated populations , animal and plant breeding , trait variation in humans , and the evolution of mating systems. In wild animal populations, evidence for inbreeding depression typically comes in the form of decreased juvenile survival. The detrimental effects of inbreeding on adult traits such as fecundity, longevity, offspring birth weight, and milk production are well known in agricultural and zoo populations, but their prevalence and magnitude in wild populations remain hitherto unclear.

There is no a priori genetic or ecological reason to expect inbreeding depression to be reduced late in the life cycle, other than the higher opportunity for selection in juveniles . The lack of widespread evidence for adult inbreeding depression could partly be explained by the general acceptance that inbred individuals may be rare, due to inbreeding avoidance , and inbred adults even rarer when there is selection against inbred juveniles . More importantly, estimating inbreeding in natural populations is not trivial.

Pedigree-based inbreeding coefficients (Fped-ped), the traditional estimate of genome-wide homozygosity through identity by descent (IBD), are not available for most natural populations, as accurate and sufficiently deep pedigrees are difficult to construct. Short or incomplete pedigrees generate downward biased estimates of inbreeding coefficients, which decreases the power to detect inbreeding depression. Bias may be reduced through exclusion of individuals with little or no ancestry information, such as founders, immigrants, and their offspring, but this can drastically reduce sample size . Consequently, relatively few studies of inbreeding depression in wild populations are based on pedigrees, although there are some notable exceptions.

As an alternative to pedigrees, average homozygosity at genetic markers has been widely used to estimate an individual's inbreeding status. However, for many natural populations, genetic resources have hitherto been limited to small panels of markers, typically microsatellites, yielding estimates of marker homozygosity that are often poorly correlated with genome-wide homozygosity. Consequently, correlations between microsatellite homozygosity and phenotypic trait values are typically small, although a metaanalysis showed an overall weak effect of decreasing fitness with increasing microsatellite homozygosity.

In recent years, genome-wide high-density marker data, such as that provided by panels of Single-Nucleotide Polymorphisms (SNPs), are becoming available for an increasing number of species. Theory and simulations predict that average homozygosity at a large number of SNPs provides a more precise estimate of genome-wide homozygosity than inbreeding coefficients from even a perfect pedigree, because it can capture the variation in IBD around the pedigree expectation, brought about by Mendelian segregation and recombination. By chance, and because of physical linkage, some individuals with the same pedigree inbreeding coefficient (e.g., Fped-ped =0.25=0.25 after a mating between full siblings) inherit a larger proportion of their alleles from the same ancestral copy than expected, including at causal loci. The performance of genomic relative to pedigree-based estimators of inbreeding and pairwise relatedness has been shown in simulations and, among others, humans and cattle. The potential of large SNP panels in the wild was recently demonstrated in a small nonpedigreed sample of harbor seals, in which SNP heterozygosity showed a strong and highly significant association with parasite burden, whereas microsatellite heterozygosity did not.

However, in general it is not fully clear how novel SNP-based metrics of inbreeding will compare with the traditional pedigree-based measures in studies of wild populations. Their performance may not be as immaculate in real datasets as in simulations, as genomic estimators, like microsatellite markers, rely on a correlation in homozygosity between loci within individuals. This so-called identity disequilibrium comes about through a fraction of systematic consanguineous matings, genetic drift, or admixture , but its magnitude can be difficult to predict.

Heterosis - manifestations of heterosis source examples

Heterosis: Manifestation and Application | Plant Breeding

Manifestation of Heterosis:

Performance or expression of any character or trait is influenced by many genetic factors — some are positive (stimulating) and others are negative (decreasing). Expressivity of the genes or the degree of manifestation of a character is the result of genetic balance in the action of differently directed factors.

The various manifestations of heterosis may be summarised as follows:

1. **Increased Yield:** Increase in yield which may be measured in terms of grain, fruit, seed, leaf, tuber or the whole plant is one of the most important manifes-tations of heterosis.
2. **Increase in Size and General Vigour:** Heterosis results in more vigorous growth which ultimately leads to healthier and faster growing plants with increase in size than the parents.
3. **Better Quality:** In many cases heterosis yields better quality which may be accompanied with higher yield.
4. **Greater Adaptability:** Hybrids are generally more adapted to environmental changes than the inbred lines due to heterozygosity.
5. **More Disease Resistant:** Heterosis sometimes results into development of more disease resistant character in the hybrids.
6. **Increased Reproductive Ability:** Hybrids exhibit heterosis by expressing high fertility rate or reproductive ability, which is ultimately expressed in yield character.
7. **Increase in Growth Rate:** In many cases the hybrids show faster growth rate than the parents, but that does not always produce larger plant size than the parents.
8. **Early Flowering and Maturity:** In many cases the hybrids may show early-ness in flowering and maturity than the parents, for some crops these are the desir-able characters for crop improvement. All these manifestations of heterosis can be traced at all levels of hybrid plant organisation.

Molecular Level:

Heterosis is manifested in increased rate of DNA reduplication, transcription and translation influencing the formation of genetic information, enzymatic activity, other regulatory mechanisms and also hybrid protein molecule formation.

Functional Level:

Heterosis is expressed as an effective regulation in metabolic processes and morphogenesis in hybrid organism.

Cellular Level:

Due to change in electro-kinetic properties of hybrid cell nuclei, the heterosis is manifested by increased mitosis.

Organism Level:

Heterosis is expressed as increased growth and differentiation of vegetative organs, synthesis and accumulation of nutritional substances and utilisation of metabolic process for yield formation.

Genetic Basis of Heterosis:

There are two main theories to explain the genetic cause of heterosis.

Dominance Hypothesis:

This hypothesis was proposed by Davenport and further expanded by others. This hypothesis suggests that at each locus dominant allele has the favourable character, whereas the recessive allele has the unfavourable character.

When they are combined together; i.e., in heterozygous condition in the hybrids, the favourable characters get expressed whereas the unfavourable characters are masked. So the heterosis results from the masking of harmful effects of recessive alleles by their dominant alleles.

Dominance Hypothesis has Assumptions:

(a) Dominant genes are beneficial and recessive genes are deleterious.

(b) The loci show addition effects, non-allelic interactions are absent.

(c) No recombination barrier between the genes.

With the help of following example heterosis can be explained:

In a cross between Inbred A (AAbbCCdd) with Inbred B (AAbbCCdd), there will be no heterosis in F_1 hybrid, there is no masking of recessive gene in hybrid. But in another cross, Inbred A (AAbbCCdd) is crossed with Inbred D (aaBBccDD), where the F_1 hybrid is (AaBbCcDd) with all the genes having dominant allele.

As a result the harmful effects of a, b, c, d are hidden by the dominant alleles A, B, C and D. Thus some parents produce heterotic progeny while others do not. Generally parents of diverse or different origin are more likely to produce heterotic progeny than those of similar origin.

Failure in Isolation of Inbreds as Vigorous as Hybrids:

According to domi-nance hypothesis it should be possible to get the inbred line with all the dominant genes. Such inbreds should be as vigorous as the F_1 hybrids, but such inbreds have not been isolated.

Symmetrical Distribution in F_2:

According to dominance hypothesis, the quan-titative characters should not show symmetrical distribution as because dominant and recessive alleles should segregate in the proportion of 3: 1, but generally the F_2 shows symmetrical distribution.

Above two objections can be explained by linked genes. Many of the quantitative characters are governed by linked genes together, so to get the inbred line with all domi-nant genes require several precisely placed crossovers. In another explanation it can be showed that if the number of genes governing the quantitative characters is large, symmetrical distribution would be obtained even without linkage.

Over-dominance Hypothesis:

This hypothesis was independently proposed by East and Shull. This is sometimes known as single gene heterosis, super-dominance, cumulative action of divergent alleles and stimulation of divergent alleles. According to this hypothesis, heterozygotes are supe-rior to both the homozygotes.

So the heterozygote Aa would be superior to both the homozygotes AA and aa. Consequently, heterozygosity is essential for the cause of heterosis. In case of maize, the gene ma affects maturity. The heterozygote Ma/ma is more vigorous with late maturity than the homozygotes Ma/Ma or ma/ma.

Another proposal by East was that there are several alleles, e.g., a_1, a_2, a_3, a_4...................... etc. with increasingly different functions. Heterozygotes between more divergent alleles would be more heterotic than those involving less divergent genes, e.g., a_1a_4 is more heterotic than a_1a_2, a_2a_3, a_3a_4, etc. In these cases due to presence of divergent alleles the hybrids have the capacity to perform different functions which is not possible by any of the heterozygotes.

Objection:

1. There are many examples where the superiority is due to the epistatic affect of several non-allelic genes, not due to over-dominance (which is the interaction between allelic genes).

2. There is another objection against over-dominance hypothesis that there are many examples where the homozygotes are superior to the heterozygotes.

Physiological Basis of Heterosis:

Hybrid vigour, the product of heterotic mechanism, is essentially a physiological manifestation.

This better physiological efficiency of hybrids is derived chiefly from:

1. Better initial growth.
2. Greater uptake followed by better utilisation of nutrients by hybrids.

The initial growth activities include the different physiological processes during ger-mination:

(a) Efficient water absorption,

(b) Better activity of enzymes,

(c) Rapid mobilization and utilization of stored food matter,

(d) Transformation and building up of active protoplasmic synthesis.

To explain all these processes different hypotheses have been put forwarded:

Initial Capital and Physiological Stimulus:

Large embryo and seed size in hybrids provide initial advantage to the hybrid during germination and early growth of seedlings. This hypothesis is debatable due to two reasons: the greater seedling vigour always not associated with maturity and also hybrid seeds are need not to be always with large size to attain hybrid vigour.

Balanced Metabolism and Heterosis at Molecular Level:

The hybrids are endowed with a more balanced metabolism than their inbred parents. Many of the enzymes of heterotic plants exhibit greater efficiency over those of their better parents. The hybrids show better and rapid unfolding of balanced meta-bolic processes.

Mitochondrial Complementation and Heterosis:

ATPase activity of the mixture of mitochondria from different inbred lines of maize sometimes ex-ceed that of the mitochondria of individual lines. This heterotic effect is called as mitochondrial complementation.

The mitochondria of heterotic hybrids ab-sorb more O_2 and have high P/O index, i.e., phosphorylation/oxidation ratio than those of inbred lines and non-heterotic hybrids. This suggests that high level of oxidising phosphorylation and synthesis of high energy ATP bonds create favourable conditions for biosynthetic processes and important requi-site for heterotic development.

Greater Ability for uptake and Utilisation of Nutrients:

Heterosis in post germination seedling growth is associated with better absorption and assimila-tion of several specific substances essential to the fundamental growth pro-cesses of the organism; such as nutritional factors, water absorption and other factors.

Efficient uptake and assimilation of nutrients by heterotic hybrid seed-lings confer the following advantages:

1. Larger number of leaf primordia.
2. High carboxylase and photophosphorylation activity.
3. Greater leaf area and larger number of leaves.
4. More branches per panicle and more grains per branch.
5. High grain weight, etc.

Applications of Heterosis in Plant Breeding:

Heterosis is observed in almost every crop species studied, the application of this phenomenon for its commercial exploitation depends on the expression of the degree of heterosis. This phenomenon is commercially used to produce hybrid or synthetic variet-ies, which needs the maximisation of its expression and also fixation.

Maximisation of hybrid vigour (HF_1) can be achieved by increasing either 'd' (direc-tional dominance) or 'y' (initial differences in gene frequency between parents), i.e., choosing genetically divergent parents, HF_1 = "dy.

Fixation of hybrid vigour is needed for its commercial application which can be done by vegetative propagation, or by stable apomictic reproduction, or by transferring heterozygosity to polyploid or fixation by obtaining structural heterozygotes.

Fixation of heterosis in crops like potato, sweet potato, sugarcane, sugarbeet and many ornamental plants can be achieved by vegetative propagation as there seeds are not essential. Incorporation of genes conditioning vegetative apospory or diploid partheno-genesis in hybrid seed crops may lead to a permanent heterozygote advantage.

Heterozy-gosity can be maintained or saved from being lost due to segregation by converting dip-loid heterozygotes into tetraploid or hexaploid. Incorporation of gaudensvelans combi-nation helps to survive only the heterozygotes not the homozygous combination — such mechanism may be introduced in crop plants, thereby hybrid vigour can be fixed with great success.

Heterosis or hybrid vigour have been commercially utilised in both cross pollinated and in some self-pollinated species. In most of the cases the utilisation of this heterosis phenomenon is not successful because of difficulty in production of large quantities of hybrid seeds. This is particularly difficult in self-pollinated species.

Heterosis

Time course imaging of two maize inbreds and their F1 hybrid (middle) exhibiting heterosis.

Heterosis, hybrid vigor, or outbreeding enhancement is the improved or increased function of any biological quality in a hybrid offspring. An offspring is heterotic if its traits are enhanced as a result of mixing the genetic contributions of its parents. The heterotic offspring often has traits that are more than the simple addition of the parents' traits, and can be explained by Mendelian or non-Mendelian inheritance. Typical heterotic/hybrid traits of interest in agriculture are higher yield, quicker maturity, stability, drought tolerance etc.

Definitions

In proposing the term **heterosis** to replace the older term **heterozygosis**, G.H. Shull aimed to avoid limiting the term to the effects that can be explained by heterozygosity in Mendelian inheritance.

The physiological vigor of an organism as manifested in its rapidity of growth, its height and general robustness, is positively correlated with the degree of dissimilarity in the gametes by whose union the organism was formed ... The more numerous the differences between the uniting gametes — at least within certain limits — the greater on the whole is the amount of stimulation ... These differences need not be Mendelian in their inheritance ... To avoid the implication that all the genotypic differences which stimulate cell-division, growth and other physiological activities of an organism are Mendelian in their inheritance and also to gain brevity of expression I suggest ... that the word 'heterosis' be adopted.

Heterosis is often discussed as the opposite of inbreeding depression, although differences in these two concepts can be seen in evolutionary considerations such as the role of genetic variation or the effects of genetic drift in small populations on these concepts. Inbreeding depression occurs when related parents have children with traits that negatively influence their fitness largely due to homozygosity. In such instances, outcrossing should result in heterosis.

Not all outcrosses result in heterosis. For example, when a hybrid inherits traits from its parents that are not fully compatible, fitness can be reduced. This is a form of outbreeding depression, the effects of which are similar to inbreeding depression.

Genetic and epigenetic bases

Since the early 1900s, two competing genetic hypotheses, not necessarily mutually exclusive, have been developed to explain hybrid vigor. More recently, an epigenetic component of hybrid vigor has also been established.

Dominance and overdominance

When a population is small or inbred, it tends to lose genetic diversity. Inbreeding depression is the loss of fitness due to loss of genetic diversity. Inbred strains tend to be homozygous for recessive alleles that are mildly harmful (or produce a trait that is undesirable from the standpoint of the breeder). Heterosis or hybrid vigor, on the other hand, is the tendency of outbred strains to exceed both inbred parents in fitness.

Selective breeding of plants and animals, including hybridization, began long before there was an understanding of underlying scientific principles. In the early 20th century, after Mendel's laws came to be understood and accepted, geneticists undertook to explain the superior vigor of many plant hybrids. Two competing hypotheses, which are not mutually exclusive, were developed:

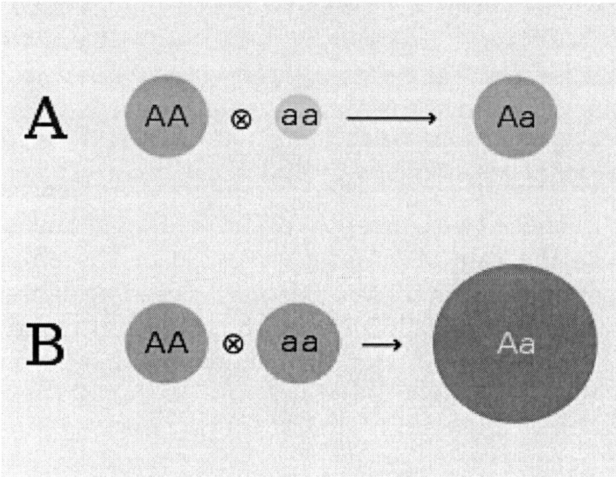

Genetic basis of heterosis. *Dominance hypothesis*. Scenario A. Fewer genes are under-expressed in the homozygous individual. Gene expression in the offspring is equal to the expression of the fittest parent. *Overdominance hypothesis*. **Scenario B**. Over-expression of certain genes in the heterozygous offspring. (The size of the circle depicts the expression level of gene A)

- **Dominance hypothesis**. The dominance hypothesis attributes the superiority of hybrids to the suppression of undesirable recessive alleles from one parent by dominant alleles from the other. It attributes the poor performance of inbred strains to loss of genetic diversity, with the strains becoming purely homozygous at many loci. The dominance hypothesis was first expressed in 1908 by the geneticist Charles Davenport. Under the dominance hypothesis, deleterious alleles are expected to be maintained in a random-mating population at a selection–mutation balance that would depend on the rate of mutation, the effect of the alleles and the degree to which alleles are expressed in heterozygotes.
- **Overdominance hypothesis**. Certain combinations of alleles that can be obtained by crossing two inbred strains are advantageous in the heterozygote. The overdominance hypothesis attributes the heterozygote advantage to the survival of many alleles that are recessive and harmful in homozygotes. It attributes the poor performance of inbred strains to a high percentage of these harmful recessives. The overdominance hypothesis was developed independently by Edward M. East

(1908) and George Shull (1908). Genetic variation at an overdominant locus is expected to be maintained by balancing selection. The high fitness of heterozygous genotypes favours the persistence of an allelic polymorphism in the population. This hypothesis is commonly invoked to explain the persistence of some alleles (most famously the Sickle cell trait allele) that are harmful in homozygotes. In normal circumstances, such harmful alleles would be removed from a population through the process of natural selection. Like the dominance hypothesis, it attributes the poor performance of inbred strains to expression of such harmful recessive alleles.

Dominance and overdominance have different consequences for the gene expression profile of the individuals. If overdominance is the main cause for the fitness advantages of heterosis, then there should be an over-expression of certain genes in the heterozygous offspring compared to the homozygous parents. On the other hand, if dominance is the cause, fewer genes should be under-expressed in the heterozygous offspring compared to the parents. Furthermore, for any given gene, the expression should be comparable to the one observed in the fitter of the two parents. In any case, outcross matings provide the benefit of masking deleterious recessive alleles in progeny.

Historical retrospective

Which of the two mechanisms are the "main" reason for heterosis has been a scientific controversy in the field of genetics. Population geneticist James Crow (1916–2012) believed, in his younger days, that overdominance was a major contributor to hybrid vigor. In 1998 he published a retrospective review of the developing science. According to Crow, the demonstration of several cases of heterozygote advantage in *Drosophila* and other organisms first caused great enthusiasm for the overdominance theory among scientists studying plant hybridization. But overdominance implies that yields on an inbred strain should decrease as inbred strains are selected for the performance of their hybrid crosses, as the proportion of harmful recessives in the inbred population rises. Over the years, experimentation in plant genetics has proven that the reverse occurs, that yields increase in both the inbred strains and the hybrids, suggesting that dominance alone may be adequate to explain the superior yield of hybrids. Only a few conclusive cases of overdominance have been reported in all of genetics. Since the 1980s, as experimental evidence has mounted, the dominance theory has made a comeback.

Crow wrote:

The current view ... is that the dominance hypothesis is the major explanation of inbreeding decline and [of] the high yield of hybrids. There is little statistical evidence for contributions from overdominance and epistasis. But whether the best hybrids are getting an extra boost from overdominance or favorable epistatic contributions remains an open question.

Epigenetics

An epigenetic contribution to heterosis has been established in plants, and it has also been reported in animals. MicroRNAs (miRNAs), discovered in 1993, are a class of non-coding small RNAs which repress the translation of messenger RNAs (mRNAs) or cause degradation of mRNAs. In hybrid plants, most miRNAs have non-additive expression (it might be higher or lower than the levels in the parents). This suggests that the small RNAs are involved in the growth, vigor and adaptation of hybrids.

'Heterosis without hybridity' effects on plant size have been demonstrated in genetically isogenic F1 triploid (autopolyploid) plants, where paternal genome excess F1 triploids display positive heterosis, whereas maternal genome excess F1s display negative heterosis effects. Such findings demonstrate that heterosis effects, with a genome dosage-dependent epigenetic basis, can be generated in F1 offspring that are genetically isogenic (i.e. harbour no heterozygosity). It has been shown that hybrid vigor in an allopolyploid hybrid of two *Arabidopsis* species was due to epigenetic control in the upstream regions of two genes, which caused major downstream alteration in chlorophyll and starch accumulation. The mechanism involves acetylation and/or methylation of specific amino acids in histone H3, a protein closely associated with DNA, which can either activate or repress associated genes.

Specific mechanisms

Major histocompatibility complex in animals

One example of where particular genes may be important in vertebrate animals for heterosis is the major histocompatibility complex (MHC). Vertebrates inherit several copies of both MHC class I and MHC class II from each parent, which are used in antigen presentation as part of the adaptive immune system. Each different copy of the genes is able to bind and present a different set of potential peptides to T-lymphocytes. These genes are highly polymorphic throughout populations, but are more similar in smaller, more closely related populations. Breeding between more genetically distant individuals decreases the chance of inheriting two alleles that are the same or similar, allowing a more diverse range of peptides to be presented. This, therefore, gives a decreased chance that any particular pathogen will not be recognised, and means that more antigenic proteins on any pathogen are likely to be recognised, giving a greater range of T-cell activation, so a greater response. This also means that the immunity acquired to the pathogen is against a greater range of antigens, meaning that the pathogen must mutate more before immunity is lost. Thus, hybrids are less likely to succumb to pathogenic disease and are more capable of fighting off infection. This may be the cause, though, of autoimmune diseases.

Plants

Crosses between inbreds from different **heterotic groups** result in vigorous F1 hybrids with significantly more heterosis than F1 hybrids from inbreds within the same heterotic group or pattern. Heterotic groups are created by plant breeders to classify inbred lines, and can be progressively improved by reciprocal recurrent selection.

Heterosis is used to increase yields, uniformity, and vigor. Hybrid breeding methods are used in maize, sorghum, rice, sugar beet, onion, spinach, sunflowers, broccoli and to create a more psychoactive cannabis.

Corn (maize)

Nearly all field corn (maize) grown in most developed nations exhibits heterosis. Modern corn hybrids substantially outyield conventional cultivars and respond better to fertilizer.

Corn heterosis was famously demonstrated in the early 20th century by George H. Shull and Edward M. East after hybrid corn was invented by Dr. William James Beal of Michigan State University based on work begun in 1879 at the urging of Charles Darwin. Dr. Beal's work led to the first published account of a field experiment demonstrating hybrid vigor in corn, by Eugene Davenport and Perry Holden, 1881. These various pioneers of botany and related fields showed that crosses of inbred lines made from a Southern dent and a Northern flint, respectively, showed substantial heterosis and outyielded conventional cultivars of that era. However, at that time such hybrids could not be economically made on a large scale for use by farmers. Donald F. Jones at the Connecticut Agricultural Experiment Station, New Haven invented the first practical method of producing a high-yielding hybrid maize in 1914–1917. Jones' method produced a double-cross hybrid, which requires two crossing steps working from four distinct original inbred lines. Later work by corn breeders produced inbred lines with sufficient vigor for practical production of a commercial hybrid in a single step, the single-cross hybrids. Single-cross hybrids are made from just two original parent inbreds. They are generally more vigorous and also more uniform than the earlier double-cross hybrids. The process of creating these hybrids often involves detasseling.

Temperate maize hybrids are derived from two main heterotic groups: 'Iowa Stiff Stalk Synthetic', and nonstiff stalk.

Rice (Oryza sativa)

Rice production has seen enormous rise in China due to heavy uses of hybrid rice. In China, efforts have generated a super hybrid rice strain ('LYP9') with a production capability around 15 tons per hectare. In India also, several varieties have shown high vigor, including 'RH-10' and 'Suruchi 5401'.

Animals

Hybrid livestock

The concept of heterosis is also applied in the production of commercial livestock. In cattle, crosses between Black Angus and Hereford produce a cross known as a "Black Baldy". In swine, "blue butts" are produced by the cross of Hampshire and Yorkshire. Other, more exotic hybrids (two different species, so genetically more dissimilar), such as "beefalo" which are hybrids of cattle and bison, are also used for specialty markets.

Poultry

Within poultry, sex-linked genes have been used to create hybrids in which males and females can be sorted at one day old by color. Specific genes used for this are genes for barring and wing feather growth. Crosses of this sort create what are sold as Black Sex-links, Red Sex-links, and various other crosses that are known by trade names.

Commercial broilers are produced by crossing different strains of White Rocks and White Cornish, the Cornish providing a large frame and the Rocks providing the fast rate of gain. The hybrid vigor produced allows the production of uniform birds at a marketable carcass weight at 6–9 weeks of age.

Likewise, hybrids between different strains of White Leghorn are used to produce laying flocks that provide the majority of white eggs for sale in the United States.

Dogs

In 2013, a study found that mixed breeds live on average 1.2 years longer than pure breeds.

John Scott and John L. Fuller performed a detailed study of purebred Cocker Spaniels, purebred Basenjis, and hybrids between them. They found that hybrids ran faster than either parent, perhaps due to heterosis. Other characteristics, such as basal heart rate, did not show any heterosis—the dog's basal heart rate was close to the average of its parents—perhaps due to the additive effects of multiple genes.

Sometimes people working on a dog-breeding program find no useful heterosis.

All this said, studies do not provide definitive proof of hybrid vigor in dogs. This is largely due to the unknown heritage of most mixed breed dogs used. Results vary wildly, with some studies showing benefit and others finding the mixed breed dogs to be more prone to genetic conditions.

Birds

In 2014, a study undertaken by the Centre for Integrative Ecology at Deakin University in Geelong, Victoria, concluded that intraspecific hybrids between the subspecies *Platycercus elegans flaveolus* and *P. e. elegans* of the crimson rosella (*P. elegans*) were more likely to fight off diseases than their pure counterparts.

Humans

Human beings are all extremely genetically similar to one another. Michael Mingroni has proposed heterosis, in the form of hybrid vigor associated with historical reductions of the levels of inbreeding, as an explanation of the Flynn effect, the steady rise in IQ test scores around the world during the 20th century, though a review of nine studies found that there is no evidence to suggest inbreeding has an effect on IQ.

Controversy

The term heterosis often causes confusion and even controversy, particularly in selective breeding of domestic animals, because it is sometimes (incorrectly) claimed that all crossbred plants and animals are "genetically superior" to their parents, due to heterosis,. but two problems exist with this claim:

1. according to an article published in the journal *Genome Biology*, "genetic superiority" is an ill-defined term and not generally accepted terminology within the scientific field of genetics. A related term fitness is well defined, but it can rarely be directly measured. Instead, scientists use objective, measurable quantities, such as the number of seeds a plant produces, the germination rate of a seed, or the percentage of organisms that survive to reproductive age. From this perspective, crossbred plants and animals exhibiting heterosis may have "superior" traits, but this does not necessarily equate to any evidence of outright "genetic superiority". Use of the term "superiority" is commonplace for example in crop breeding, where it is well understood to mean a better-yielding, more robust plant for agriculture. Such a plant may yield better on a farm, but would likely struggle to survive in the wild, making this use open to misinterpretation. In human genetics any question of "genetic superiority" is even more problematic due to the historical and political implications of any such claim. Some may even go as far as to describe it as a questionable value judgement in the realm of politics, not science.
2. not all hybrids exhibit heterosis.

   An example of the ambiguous value judgements imposed on hybrids and hybrid vigor is the mule. While mules are almost always infertile, they are valued for a combination of hardiness and temperament that is different from either of their horse or donkey parents. While these qualities may make them "superior" for particular uses by humans, the infertility issue implies that these animals would most likely become extinct without the intervention of humans through animal husbandry, making them "inferior" in terms of natural selection.

Genetics basis of inbreeding depression & heterosis classical view of genetic basis of heterosis

Data on the effects of inbreeding on fitness components are reviewed in the light of population genetic models of the possible genetic causes of inbreeding depression. Deleterious mutations probably play a major role in causing inbreeding depression. Putting together the different kinds of quantitative genetic data, it is difficult to account for the very large effects of inbreeding on fitness in *Drosophila* and outcrossing plants without a significant contribution from variability maintained by selection. Overdominant effects of alleles on fitness components seem not to be important in most cases.

Inbreeding depression

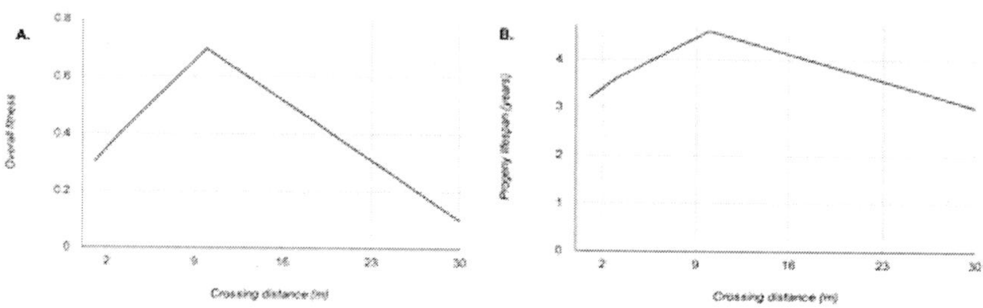

Inbreeding depression in *Delphinium nelsonii*. **A.** Overall fitness of progeny cohorts and the **B.** progeny lifespan were all lower when progeny were the result of crosses with pollen taken close to a receptor plant.

Inbreeding depression is the reduced biological fitness which has the potential to result from inbreeding (the breeding of related individuals). Biological fitness refers to an organism's ability to survive and perpetuate its genetic material. Inbreeding depression is often the result of a population bottleneck. In general, the higher the genetic variation or gene pool within a breeding population, the less likely it is to suffer from inbreeding depression, though inbreeding and outbreeding depression can simultaneously occur.

Inbreeding depression seems to be present in most groups of organisms, but varies across mating systems. Hermaphroditic species often exhibit lower degrees of inbreeding depression than outcrossing species, as repeated generations of selfing is thought to purge deleterious alleles from populations. For example, the outcrossing nematode (roundworm) *Caenorhabditis remanei* has been demonstrated to suffer severely from inbreeding depression, unlike its hermaphroditic relative *C. elegans*, which experiences outbreeding depression.

Mechanisms

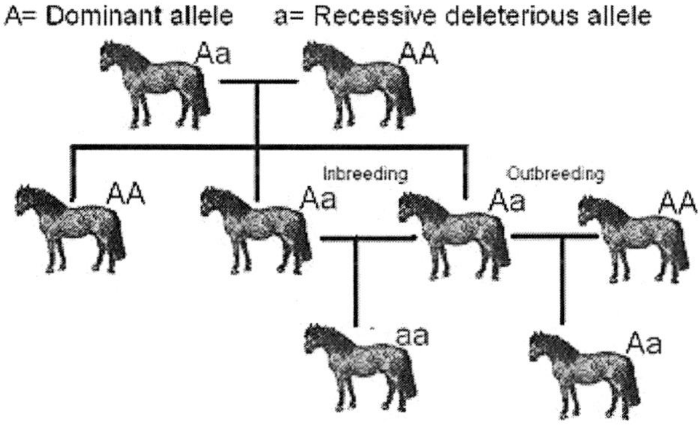

Example of inbreeding depression

Inbreeding (i.e., breeding between closely related individuals) results in more recessive traits manifesting themselves, as the genomes of pair-mates are more similar. Recessive traits can only occur in an offspring if present in both parents' genomes. The more genetically similar the parents are, the more often recessive traits appear in their offspring. This normally has a positive effect, as most genes are undergoing purifying selection (the homozygous state is favored). However, for very closely related individuals, there is an increased likelihood of homozygous deleterious genes in the offspring which can result in unfit individuals. For the alleles that confer an advantage in the heterozygous and/or homozygous-dominant state, the fitness of the homozygous-recessive state may even be zero (meaning sterile or unviable offspring).

An example of inbreeding depression is shown to the right. In this case, *a* is the recessive allele which has negative effects. In order for the *a* phenotype to become active, the gene must end up as homozygous *aa* because in the geneotype A*a*, the A takes dominance over the *a* and the *a* does not have any effect. Some recessive genes result in detrimental phenotypes by causing the organism to be less fit to its natural environment.

Another mechanism responsible for inbreeding depression is the fitness advantage of heterozygosity, which is known as overdominance. This can lead to reduced fitness of a population with many homozygous genotypes, even if they are not deleterious or recessive. Here, even the dominant alleles result in reduced fitness if present homozygously.

Overdominance is rare in nature. For practical applications, e.g. in livestock breeding, the former is thought to be more significant – it may yield completely unviable offspring (meaning outright failure of a pedigree), while the latter can only result in relatively reduced fitness.

Natural selection

Natural selection cannot effectively remove all deleterious recessive genes from a population for several reasons. First, deleterious genes arise constantly through de novo mutation within a population. Second, most offspring will have some deleterious traits, so few will be more fit for survival than the others. Different deleterious traits are extremely unlikely to equally affect reproduction – an especially disadvantageous recessive trait expressed in a homozygous recessive individual is likely to eliminate itself, naturally limiting the expression of its phenotype. Third, recessive deleterious alleles will be "masked" by heterozygosity, and so in a dominant-recessive trait, heterozygotes will not be selected against.

When recessive deleterious alleles occur in the heterozygous state, where their potentially deleterious expression is masked by the corresponding wild-type allele, this masking phenomenon is referred to as complementation.

In general, sexual reproduction in eukaryotes has two fundamental aspects: genetic recombination during meiosis, and outcrossing. It has been proposed that these two aspects have two natural selective advantages respectively. A proposed adaptive advantage of meiosis is that it facilitates recombinational repair of DNA damages that are otherwise difficult to repair. A proposed adaptive advantage of outcrossing is complementation, which is the masking of deleterious recessive alleles. The selective advantage of complementation may largely account for avoidance of inbreeding, though it is unlikely that animals avoid inbreeding.

Management

Hybridization as a conservation effort is be appropriate if the population has lost "substantial genetic variation through genetic drift and the detrimental effects of inbreeding depression are apparent" and a similar population should be used. Different populations of the same species have different deleterious traits, and therefore their cross breeding is less likely to result in homozygosity at most loci in the offspring. This is known as outbreeding enhancement, which can be performed in extreme cases of severe inbreeding by conservation managers and zoo captive breeders to prevent inbreeding depression.

However, intermixing two different populations can give rise to unfit polygenic traits in outbreeding depression (i.e. yielding offspring which lack the genetic adaptations to specific environmental conditions). These, then, will have a lowered fitness than pure-bred individuals of a particular subspecies that has adapted to its local environment.

In humans

Inbreeding may have both detrimental and beneficial effects. The biological effects of inbreeding depression in humans can on occasion be confounded by socioeconomic and cultural influences on reproductive behavior. Studies in human populations have shown that age at marriage, duration of marriage, illiteracy, contraceptive use, and reproductive compensation are the major determinants of apparent fertility, even amongst populations with a high proportion of consanguinous unions. However, several small effects on increased mortality, longer inter-birth intervals and reduced overall productivity have been noted in certain isolated populations, though other studies show increased fitness of offspring and no effect on lifespan past the 2nd cousin level.

Charles Darwin was one of the first scientists to demonstrate the effects of inbreeding depression, through numerous experiments on plants. Darwin's wife, Emma, was his first cousin, and he was concerned about the impact of inbreeding on his ten children, three of whom died at age ten or younger; three others had childless long-term marriages.

Humans do not seek to completely minimize inbreeding, but rather to maintain an optimal amount of inbreeding vs. outbreeding. Close inbreeding reduces fitness through inbreeding depression, but some inbreeding brings benefits. Indeed, inbreeding "increases the speed of selection of beneficial recessive and co-dominant alleles, e.g. those that protect against diseases."

Factors reducing inbreeding depression

Whilst inbreeding depression has been found to occur in almost all sufficiently studied species, some taxa, most notably some angiosperms, appear to suffer lower fitness costs than others in inbred populations. Three mechanisms appear to be responsible for this: purging, differences in ploidy, and selection for heterozygosity. It must be cautioned that some studies failing to show an absence of inbreeding depression in certain species can arise from small sample sizes or where the supposedly outbred control group is already suffering inbreeding depression, which frequently occurs in populations that have undergone a recent bottleneck, such as those of the naked mole rat.

Purging selection

Purging selection occurs where the phenotypes of deleterious recessive alleles are exposed through inbreeding, and thus can be selected against. This can lead to such detrimental mutations being removed from the population, and has been demonstrated to occur rapidly where the recessive alleles have a lethal effect. The efficiency of purging will depend on the relationship between the magnitude of the deleterious effect that is unmasked in the homozygotes and the importance of genetic drift, so that purging is weaker for non-lethal than for recessive lethal alleles. For very small populations, drift has a strong influence, which can cause the fixation of sublethal alleles under weak selection. The fixation of a single allele for a specific gene can also reduce fitness where heterozygote advantage was previously present (i.e., where heterozygous individuals have higher fitness than homozygotes of either allele), although this phenomenon seems to make a usually small contribution to inbreeding depression. Although naturally occurring, purging can be important for population survival, deliberately attempting to purge deleterious mutations from a population is not generally recommended as a technique to improve the fitness of captive bred animals. In plants, genetic load can be assessed through a test analogous to an inbreeding depression test called an Autogamy depression test.

Polyploidy

Many angiosperms (flowering plants) can self-fertilise for several generations and suffer little from inbreeding depression. This is very useful for species which disperse widely and can therefore find themselves growing in a novel environment with no conspecifics present. Polyploidy (having more than two paired sets of each chromosome), which is prevalent in angiosperms, ferns and a select few animal taxa, accounts for this. By having several copies of a chromosome, as opposed to two, homozygosity is less likely to occur in inbred offspring. This means that recessive deleterious alleles are not expressed as frequently as with many copies of a chromosome; it is more likely that at least one will contain a functional allele.

Selection for heterozygosity

Selection for heterozygosity is rare, as lost loci undergo purifying selection for homozygous loci. Inbreeding depression has also been found to occur more gradually than predicted in some wild populations, such as in the highly inbred population of Scandinavian wolves. This appears to be due to a selection pressure for more heterozygous individuals, which generally are in better condition and so are more likely to become one of the few animals to breed and produce offspring.

Genetic basis of heterosis

Two competing hypotheses, not necessarily mutually exclusive, have been developed to explain hybrid vigor. The **dominance hypothesis** attributes the superiority of hybrids to the suppression of undesirable (deleterious) recessive alleles from one parent by dominant alleles from the other. It attributes the poor performance of inbred strains to the loss of genetic diversity, with the strains becoming purely homozygous deleterious alleles at many loci. The **overdominance hypothesis** states that some combinations of alleles (which can be obtained by crossing two inbred strains) are especially advantageous when paired in a heterozygous individual. The concept of heterozygote advantage/overdominance is not restricted to hybrid lineages. This hypothesis is commonly invoked to explain the persistence of many alleles (most famously the erythrocyte-sickling allele) that are harmful in homozygotes; in normal circumstances, such harmful alleles would be removed from a population through the process of natural selection. Like the dominance hypothesis, it attributes the poor performance of many inbred strains to a high frequency of these harmful recessive alleles and the associated high frequency of homozygous-recessive genotypes.

Hybrid livestock

The concept of heterosis is also applied in the production of commercial livestock. In cattle, hybrids between Black Angus and Hereford produce a hybrid known as a "Black Baldy". In swine, "blue butts" are produced by the cross of Hampshire and Yorkshire. Other, more exotic hybrids such as "beefalo" are also used for specialty markets.

Within poultry, sex-linked genes have been used to create hybrids in which males and females can be sorted at one day old by color. Specific genes used for this are genes for barring and wing feather growth. Crosses of this sort create what are sold as Black Sex-links, Red Sex-links, and various other crosses that are known by trade names.

Commercial broilers are produced by crossing different strains of White Rocks and White Cornish, the Cornish providing a large frame and the Rocks providing the fast rate of gain. The hybrid vigor produced allows the production of uniform birds with a marketable carcass at 6–9 weeks of age.

Likewise, hybrids between different strains of White Leghorn are used to produce laying flocks that provide the majority white eggs for sale in the United States.

Humans

Experimental breeding of humans is considered unethical, so any evidence of heterosis in humans is derived from observational studies It has been suggested that many beneficial effects on average health, intelligence and height have resulted from an increased heterosis, in turn resulting from increased mixing of the human population such as by urbanization.

Bibliography

Ausubel, F.M. : *Current Protocols in Molecular Biology,* New York: John Wiley and Sons, 2009.

Bast, R.C. Jr.: *Cancer Medicine*, Decker, Hamilton, 2000.

Bauer, M.W., and G. Gaskell : *Biotechnology: The Making of a Global Controversy*, Cambridge: Cambridge University Press, 2002.

Bhatia, A L : *Biochemistry and Endocrinology,* Indus Valley, Delhi, 2012.

Bradley, W.G.: *Neurology in Clinical Practice*, Butterworth Heinemann, Boston, 2010.

Brookes, Martin: *Get a Grip on Genetics, Time Life Books*, East Sussex, England, 2014.

Cecilia Saccone: *Handbook of Comparative Genomics: Principles and Methodology*. New York: Wiley-Liss. 2013.

Chakraborty, Chiranjib : *Advances in Biochemistry and Biotechnology,* Daya, Delhi, 2005.

Chambers, K. L.: *Biochemical Coevolution: Eugene,* Oregon State University Press, 2010.

Clark, J.M. and Switzer, R.L. : *Experimental Biochemistry,* New York: W.H. Freeman and Company, 2013.

Diaz, E.: *Microbial Biodegradation: Genomics and Molecular Biology,* Caister Academic Press, UK, 2008.

Dobzhanshy, T.: *Genetics of the Evolutionary Process*: New York, Columbia University Press, 2010.

Flores E,: *The Cyanobacteria: Molecular Biology, Genomics and Evolution* (1st ed.). Caister Academic Press, 2008.

Friedberg, E.C.: *DNA Repair*, New York, WH Freeman and Company, 2013.

Griffiths A.J.F.: *Introduction to Genetic Analysis*, W.H. Freeman and Company, New York, USA, 2005.

Hames, B. D.: *Genes and Embryos*, Oxford University Press, New York, 2011.

Harris, John : *Wonderwoman and Superman: The Ethics of Human Biotechnology,* Oxford: Oxford University Press, 2012.

Hartl D, Jones E: *Genetics: Analysis of Genes and Genomes,* Jones & Bartlett, 2005.

Hartl, Daniel L.: *Basic Genetics*, Jones and Bartlett Publishers, Boston, 2011.

Jangir, M.L. : *Cell Biology : Fundamentals and Applications*, Agrobios, Delhi, 2009.

Klotz, J. W.: *Genes, Genesis, and Evolution*: St. Louis, Mo., Concordia Publishing Co., 2010.

Knudtson, P.: *Genethics: The Clash Between the New Genetics and Human Values*: Cambridge, Mass., Harvard University Press, 2009.

Lewontin, R.C.: *The Doctrine of DNA: Biology as Ideology*. Harmondsworth: Penguin, 2013.

Lindee, M.S. : *The DNA Mystique: The Gene as a Cultural Icon*. New York: W.H. Freeman, 2015.

Mayr, E.: *Change of Genetic Environment and Evolution*, Allen and Unwin, London, 2004.

Meffe, G.K. : *Principles of Conservation Biology,* Sunderland, MA: Sinauer Associates, 2006.

Michael R.: Cummings. *Essentials of Genetics*, Prentice Hall, New Jersey, 2014.

Migloni, G.S.: *Dictionary of Plant Genetics and Molecular Biology*, Hawthorne Press, New York, 2013.

Nisha Khalsa: *Essentials of Biochemistry*, Aavishkar Pub, Delhi, 2008.

Prasad, S.K. : *Biochemistry of Carbohydrates,* Discovery Pub, Delhi, 2010.

Raghunath Narvekar: *Handbook of Biochemistry*, Adhyayan, Delhi, 2008.

Scriver, C.R.: *The Metabolic and Molecular Basis of Inherited Disease*, McGraw Hill, New York, 2011.

Sharma, Anil : *Cell Biology, Genetics and Plant Breeding*, Indus Valley Pub, Delhi, 2006.

Singh, Bimla: *Biochemistry and Genomic Revolution*, Vista International Pub, Delhi, 2006.

Stansfield, W.D.: *A Dictionary of Genetics*, Oxford Univ. Press, New York, 2012.

Steen, R.G.: *DNA and Destiny: Nature and Nurture in Human Behaviour*. New York and London: Plenum Press, 2015.

Sturtevant, Alfred: *History of Genetics*, Harper and Row, New York, 2005.

Stwertka, Eve : *Genetic Engineering*, Franklin Watts, New York, 2012.

Stwertka, Eve and Albert: *Genetic Engineering*, Franklin Watts, New York, 2012.

Vasudev Bhatnagar: *Cell Science and Technology*, Campus Books, Delhi, 2009.

William, Kovacs D.: *An Introduction to Geotechnical Engineering*, Prentice Hall, UK, 2011.

Williams, J.G. and R.K. Patient : *Genetic Engineering*. Oxford, England: IRL Press, 2008.